ACS SYMPOSIUM SERIES **812**

Synthetic Macromolecules with Higher Structural Order

Ishrat M. Khan, EDITOR

Clark Atlanta University

American Chemical Society, Washington, DC

Library of Congress Cataloging-in-Publication Data

Synthetic macromolecules with higher structural order / Ishrat M. Khan, editor.

    p. cm.—(ACS symposium series, ISSN 0097-6156 ; 812)

    Includes bibliographical references and index.

    ISBN 0–8412–3728–X (alk. paper)

    1. Macromolecules—Congresses. I. Khan, Ishrat, 1956 – II. American Chemical Society. Division of Polymer Chemistry. III. American Chemical Society. Meeting (219th: 2000 : San Francisco, Calif.). IV. Series.

QD380.S99   2001
547'.7—dc21   2001053839

The paper used in this publication meets the minimum requirements of American National Standard for Information Sciences—Permanence of Paper for Printed Library Materials, ANSI Z39.48–1984.

PRINTED IN THE UNITED STATES OF AMERICA

Foreword

The ACS Symposium Series was first published in 1974 to provide a mechanism for publishing symposia quickly in book form. The purpose of the series is to publish timely, comprehensive books developed from ACS sponsored symposia based on current scientific research. Occasionally, books are developed from symposia sponsored by other organizations when the topic is of keen interest to the chemistry audience.

Before agreeing to publish a book, the proposed table of contents is reviewed for appropriate and comprehensive coverage and for interest to the audience. Some papers may be excluded to better focus the book; others may be added to provide comprehensiveness. When appropriate, overview or introductory chapters are added. Drafts of chapters are peer-reviewed prior to final acceptance or rejection, and manuscripts are prepared in camera-ready format.

As a rule, only original research papers and original review papers are included in the volumes. Verbatim reproductions of previously published papers are not accepted.

ACS Books Department

Contents

Higher Ordering in Synthetic Polymers

Synthetic Peptides

Macromolecular Assemblies

Preface

Synthetic macromolecules with higher structural order are being developed as new generation materials for application in synthetic enzymes, asymmetric reagents, catalysts, electronic switches and biosensors, tissue engineering, peptide drugs, and responsive drug delivery systems. The highly interdisciplinary area includes organic and polymer synthesis, peptide design and synthesis, self-assembled macromolecules, computational studies to complement spectroscopic higher structural order determination, interaction of conformationally (or higher structurally) organized oligomers and polymers with biological systems. This book is based on an American Chemical Society (ACS) Division of Polymer Chemistry, Inc. symposium held at the 219th ACS National Meeting in San Francisco, California, March 26–30, 2000. The goal of the symposium was targeting presentations of different subtopics to demonstrate the interdisciplinary nature of the area and to encourage collaboration among researchers in the continuing development of synthetic macromolecules with higher structural order.

The potential application of non-natural macromolecular systems with almost biospecific type functional properties is fascinating and will require that researchers in the complementing subtopic areas work and learn together. The book is not encyclopedic but combines complementing areas of synthetic peptides, macromolecular assemblies, and synthetic polymers to give an overall perspective. The research reported in the chapters set the stage for designing and starting innovative studies in synthetic macromolecular systems with secondary and tertiary structural order. The book *Synthetic Macromolecules with Higher Structural Order* will serve as an excellent reference book and as a guide to new comers and established scientists in this interdisciplinary research area. The book will be of interest to a diverse group of scientist including organic and physical chemists, peptide chemists, polymer chemists, biochemists, molecular biologists, and molecular scientists.

It is a pleasure to express my gratitude to the ACS Division of Polymer Chemistry, Inc. Programming Committee for approving and scheduling the symposium in San Francisco in 2000. The funding support of the symposium was provided by the ACS Division of Polymer Chemistry, Inc. and the ACS PRF Grant # 35812–SE. I greatly appreciate the help of Kelly Dennis and Stacy VanDerWall in acquisitions and Margaret Brown in editorial and production of the ACS Books Department for providing support and guidance to bring this book to fruitful completion. Thanks also to Anne Wilson for supporting the idea for a book on this topic and getting me started. Finally, sincere appreciation and thanks to the contributors for their timely efforts and for sharing their valuable research results in this forum.

Ishrat M. Khan
Department of Chemistry
Clark Atlanta University
Atlanta, GA 30314

Chapter 1

Synthetic Macromolecules with Higher Structural Order: An Overview

Ishrat M. Khan

Department of Chemistry, Clark Atlanta University, 223 James P. Brawley Drive, S.W., Atlanta, GA 30314

To develop a new generation of biomaterials for any number of medicinal/therapeutic/biomimetic applications, it is necessary to synthesize and characterize macromolecules with higher structural ordering (secondary, tertiary and quaternary). In this overview chapter, the approaches and successes in preparing secondary helical structures using non-biological building blocks are reviewed. Stable helical poly(triphenylmethylmethacrylate)s have been prepared by helix sense selective polymerization using chiral anionic initiators. A number of β- and γ-oligopeptides, from non-natural amino acids, form stable helical conformations. Helical polymers have also been prepared by helicity induction of achiral polymers via chiral acid-base interactions. Additionally, successful approaches for preparing dissymmetric macromolecular superstructures (tertiary organization) from macromolecular systems are discussed.

One of the four subunits which make up the enzyme 4-hydroxybenzoyl-CoA Thioesterase from Psuedomonas sp. Strain CBS-3 is shown in Figure 1 *(1,2)*. The three-dimensional or tertiary structure of this subunit is characterized by secondary structures consisting of a five-stranded anti-parallel β-sheet and three α-helices. The overall three-dimensional structure of any enzyme is instrumental in its functional specificity. To develop structures with functional specificities approaching those of enzymes, it is necessary for synthetic scientists to put together organized chiral secondary structures into a desired dissymmetric tertiary structure. There is significant need to develop a fundamental understanding of what is takes to readily (relatively easily) prepare chiral secondary structures and how the secondary structures may be combined into larger dissymmetric structures. Once the preparation and characterization methods of intricate three-dimensional chiral structures are available, a leap may be made from structure preparation to developing functional properties. The goal will next become developing macromolecular systems with functional specificities approaching those of biological systems. In this overview, the approaches and successes in preparing helical secondary structures will be covered, followed by successful examples of generating dissymmetric macromolecular systems.

Discussion

The approach to develop secondary structures may be divided into two categories: (a) secondary structures based on synthetic macromolecules e.g poly(triphenylmethylmethacrylate), poly(isocyanates) and (b) helical synthetic peptides or peptides from non-natural β- and γ-amino acids. This overview will focus on category 'a' type of helical structures and interested readers are encouraged to look at the excellent review of Nowick for additional information on synthetic peptides *(3)*. Pioneering work in helical synthetic oligopeptides has been independently reported by the groups of Seebach, Gellman and Hannessian. In 1996, Seebach, reported the formation of stable helical conformations with the hexapeptide H(-β-HVal-β-HAla-β-HLeu)$_2$-OH. The helical conformation is stabilized via intramolecular hydrogen bonding similar to the natural peptides from α-amino acids. NMR studies in C_5D_5N have shown that this peptide forms a left-handed 3_1 helical conformation *(4)*. In 1998, Seebach's group extended their studies and reported the preparation and characterization of helical γ-peptides; the helical conformation again is stabilized via intramolecular hydrogen bonding *(5)*. The structures for two of Seebach's non-natural amino acids are shown in Figure 2.

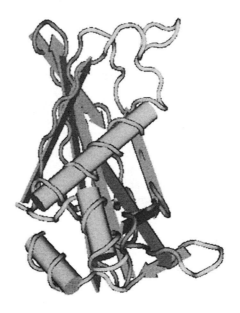

Figure 1: One of the four subunits of the enzyme 4-hyroxybenzoyl-CoA Thioesterase (2).

4

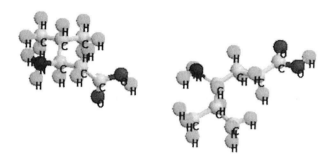

Figure 2: Non-natural β- and γ-amino acids, respectively (left to right)

Gellman's oligopeptides have been derived from optically active trans-2-aminocyclohexanecarboxylic acid. Using NMR spectroscopy, CD spectroscopy, and X-ray crystallography, Gellman and coworkers have shown that the oligopeptides form stable helical conformations in solution and in the solid state *(6,7)*. At the current time the formation of well-defined folded oligopeptides is reasonably well established, the excitement and challenge will be applying the current understanding to the preparation of larger (molecular weight) synthetic peptides with higher ordered (tertiary and quaternary) organizations.

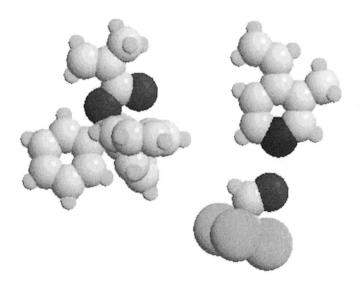

Figure 3: Space filling representation of (clockwise from left) triphenylmethylmethacrylate, 3-methy-4-vinylpyridine and chloral (8).

The Okamoto group may be considered as pioneers in the development of helical macromolecules based on vinyl monomer systems *(9)*. The group has made significant contributions in the area, starting with the successful demonstration of helix-sense selective polymerization of achiral vinyl monomers using chiral initiating complexes. For example, stable left- and right-handed helical poly(triphenylmethylmethacrylate)s have been prepared using enantiomeric chiral anionic initiators. The helical conformation of isotactic poly(triphenylmethylmethacrylate), prepared by anionic living polymerization, is the preferred low energy conformation for the polymer chain. Furthermore, the bulky triphenyl groups, because of steric factors, inhibit helix-to-helix interconversions, the helical conformation once formed is stable. Optically active helical poly(chloral) has been reported and the formation of the helix is due to steric factors favoring a particular conformation *(10)*. In Figure 3, space filling representations of the monomers, triphenylmethylmethacrylate and chloral, are shown *(8)*. The bulkiness of both the monomers is evident from the space filling representations. Steric factors are important in stabilizing the helical conformations of non-peptide based synthetic macromolecules.

In a 1958 paper, Nobel Laureate Natta showed that ortho substituted isotactic polystyrenes existed in favored helical conformations in the solid state. The ortho group is key in the steric stabilization of the helical conformation *(11)*. Therefore, the possibility existed for preparing helical polymers using sterically less demanding vinyl monomers. In 1998, the helix sense selective polymerization of 3-methyl-4-vinylpyridine was reported *(12)*. The optically active helical polymers are stable at low temperatures in solution and at room temperature in the solid state. In figure 3, the space filling representation of this monomer is compared with the bulkier triphenylmethylmethacrylate and chloral monomers. The representation shows that 3-methyl-4-vinylpyridine is sterically significantly less encumbered. Recently, Okamoto has reported the preparation of a higher structurally ordered polymer using an ortho substituted styrene, 2-[(S)-2-(1-pyrrodinylmethyl)-1-pyrrodinylmethyl]styrene *(13)*. These results suggest that preparing conformationally ordered systems from fairly simple vinyl polymers is a real possibility. Poly(isocyanates) are also polymers which because of steric factors form stable helical conformations in the solid state. In solution, the helical conformations are in a dynamic equilibrium but the equilibrium may be readily shifted by using a small excess of a chiral side group *(14-16)*. Asymmetric helix sense selective polymerization of aromatic isocyanates has been reported, the helical sense is controlled by introducing a chiral group at the α-end during initiation of the polymerization *(17)*.

Helical polymers have also been prepared by helicity induction of achiral polymer chains via acid-base interactions *(19-21)*. Stereoregular (*cis-transoidal*) poly(acetylene)s bearing acid and base functional groups may be induced into a stable helical conformation via reaction with chiral bases and acids. The formation of the conformationally stable helical structure has been demonstrated by ICD (induced circular dichroism) spectra. Induced helical structures have also been prepared from poly(phosphazenes), poly(3-methyl-4-vinylpyridine) and poly(3-carboxylphenylisocyanate) *(22-24)*. Recently, an achiral nonapeptide chain has been induced into a one-handed helical conformation via an interaction of the N-terminal amino group with chiral carboxylic acids *(25)*. The nonapeptide is achiral and before induction is present in equal concentrations of the left and the right-handed helices. The addition of a chiral acid and the interaction of the acid with the N-terminal amino group results in the formation of a predominance of a one-handed screw sense. Non-covalent interactions have been utilized to form induced helical structures from random conformations. For example the solvent induced (solvophobic) transition of the disordered conformation of a chiral(binaphtol)oligo(phenylene ethylene) in a good solvent to the compact helical conformation in a poor solvent has been reported *(26)*. Like nature, synthetic scientists must learn to manipulate the many non-covalent types of interactions to form stable higher structurally ordered macromolecules or macromolecular systems e.g. a balancing of the hydrophobic and hydrophilic forces to prepare self-organized assemblies with higher structural order. The groups of Nolte and Meijer have successfully used this approach. Helical superstructures by self-assembling of charged poly(styrene)-block-poly(isocynanodipeptide) have been obtained *(27)*. The chiral poly(isocyanodipeptide) block forms a secondary helical conformation and induces the formation of the helical superstructure. The morphology of the superstructure may be varied by adjusting parameters such as pH and individual block lengths. Interestingly enough, the sense of the helical superstructure is opposite to the sense of the poly(isocyanodipeptide) helix. Lehn has developed methods to prepare controlled supra-molecular architectures using non-covalent interactions such as hydrogen bonding and such methods remain extremely attractive to develop intricate chiral superstructures starting with optically active helical polymers *(28)*. Meijer utilized both hydrogen bonding and hydrophobic interactions to prepare helical self-assembled macromolecules *(29)*. In addition to the development of facile methods to prepare macromolecular secondary structures, self-assembling methods must also be utilized to prepare intricate chiral superstructures. Approaches are currently being developed to prepare stable complexes of peptides (secondary structures) and enzymes with biocompatible polymers in aqueous media. The information from such systems may provide ideas for developing newer methods for preparing chiral macromolecular superstructures *(30-32)*.

Conclusion

The aesthetic beauty of synthetic macromolecules with higher structural order or well defined three-dimensional structures or conformationally ordered structures is a sufficient and compelling enough reason to keep exploring this area. The potential of developing non-natural chiral macromolecular systems for application with almost any biospecific type of functional properties is fascinating. While the goal is fascinating, it remains challenging. An interdisciplinary effort including organic synthesis, peptide design and synthesis, polymer design and synthesis, computational studies to complement spectroscopic higher structural order determination, understanding the interaction of conformationally (or higher structurally ordered) organized oligomers and polymers with biological systems will be required to deliver on the promise of synthetic systems with near biological type of functional capabilities.

Acknowledgement: The author thanks National Institute of Health for research support through NIH/NIGMS/MBRS/SCORE Grant #SO6GM08247, RCMI Grant #G12RR03062 and NIH/MIRT # 5T37TW009.

References and Notes

1. Benning, M.W.; Wesenberg, G.; Liu, R.; Taylor, K.L.; Dunaway-Mariano, D.; Holden, H.M. *J. Biol. Chem.* **1998**, *273*, 33572-35579.

2. The enzyme subunit structure has been obtained using PUBMED of NCBI (National Center of Biotechnology).

3. Stigers, K.D.; Soth, M.J.; Nowick, J.J. *Curr. Opin. Chem. Bio.* **1999**, *3*, 714-723.

4. Seebach, D.; Ciceri, P.E.; Overhand, M.; Jaun, B.; Rigo, D. *Helv. Chim. Acta.* **1996**, *70*, 2043-2066.

5. Seebach, D.: Abele, S.; Gademann, K.; Guichard, G.; Hintermann, T.; Jaun, B.; Mathews, J.L.; Schreiber, J.V.; Hommel, U.; Widmer, H.; *Helv. Chem. Acta.* **1998**, 81, 932-982.

6. Appella, D.H.; Christianson, L.A.; Karle, I.L.; Powell, D.R.; Gellman, S.H. *J. Am. Chem. Soc.* **1996**, *118*, 13071-13072.

7. Appella, D.H.; Christianson, L.A.; Karle, I.L.; Powell, D.R.; Gellman, S.H. *J. Am. Chem. Soc.* **1999**, *121*, 6206-6212.

8. Space-filling representations have been generated by Alchemy 32 Version 2.05, Tripos, Inc. St. Louis, MO, USA.

9. Okamoto, Y.; Nakano, T. *Chem. Rev.* **1994**, *94*, 349-372.

10. Vogl, O.; Jaycox, G.D.; Kratky, C.; Simonsick, W.J.; Hatada, K. *Accts. Chem. Res.* **1992**, *25*, 408-413.

11. Natta, G.; Danusson, F.; Sianesi, D.; *Makromol. Chem.* **1958**, *28*, 253.

12. Ortiz, L.J.; Khan, I.M. *Macromolecules* **1998**, *31*, 5927-5929.

8

13. Habaue, S.; Ajiro, H.; Okamoto, Y.; *J. Polym. Sci. Chem. Ed.* **2000**, *38*, 4088-4094.

14. Green, M.M.; Reidy, M.P.; Johnson, R.J.; Darling, G.; O'Leary, D.J., Wilson, G. *J. Am Chem. Soc.* **1989**, *111*, 6452-6454.

15. Green, M.M.; Peterson, N.C.; Sato, T.; Teramoto, A.; Cook, R.; Lifson, S. *Science* **1995**, 268, 1860-1866.

16. Jha, S.K.; Cheon, K.S.; Green, M.M.; Selinger, J.V.; *J. Am Chem. Soc.* **1999**, *121*, 1665-1673.

17. Mayer, S.; Zentel, R. *Macromol. Chem. Phys.* **1998**, *199*, 1675-1682.

18. Okamoto, Y.; Matsuda, M.; Nakano, T.; Yashima, E.; *J. Polym. Sci. Chem. Ed.* **1994**, *32*, 309-315.

19. Yashima, E.; Maeda, K.; Okamoto, Y. *J. Am. Chem. Soc.* **1997**, *119*, 6345-6359.

20. Yashima, E.; Maeda, K.; Okamoto, Y. *Nature* **1999**, 399, 449-451.

21. Yashima, E.; Maeda, K.; Matsushima, T.; Okamoto, Y. *Chirality* **1997**, *9*, 593-600.

22. Yashima, E., Maeda, K. see Chapter in this volume.

23. Ortiz, L.J.; Pratt, L. M.; Smitherman, K.; Sannigrahi, B.; Khan, I.M., see Chapter in this volume.

24. Maeda, K.; Yamamoto, N.; Okamoto, Y. *Macromolecules* **1998**, *31*, 5924-5926.

25. Inai, Y.; Tagawa, K.; Takasu, A.; Hirabayashi, T.; Oshikawa, T.; Yamashita, M. *J. Am. Chem. Soc.* **2000**, *122*, 11731-11732.

26. Gin, M.S.; Yokozawa, T.; Prince, R.B.; Moore, J.S. *J. Am. Chem. Soc. 1999*, 121, 2643-2644.

27. Cornelissen, J.L.M.; Fischer, M.; Sommerdijk, N.A.J.M; Nolte, R.J.M. *Science* **1998**, *280*, 1427-1430.

28. Lehn, J.M. *Makromol. Chem. Macromol. Symp.* **1993**, *60*, 1-17.

29. Ky Hirschberg, J.H.K.; Brunweld, L.; Ramzi, A.; Vekemans, J.A.J.M.; Sijbesma, R.P.; Meijer, E,W. *Nature* **2000**, *407*, 167-170.

30. Pemawansa, K.P; Thakur, A.; Karikari, E.K.; Khan, I.M. *Macromolecules* **1999**, *32*, 1910-1917.

31. Thunemann, A.F.; Beyermann, Kukula, H. *Macromolecules* **2000**, *33*, 5906-5911.

32. Topchieva, I.N.; Sorokina, E.M.; Efremova, N.V.; Ksenofontov, A.L.; Kurganov, B.I. *Bioconjugate Chem.* **2000**, *11*, 22-29.

Higher Ordering in Synthetic Polymers

Chapter 2

Synthesis of Single-Handed Helical Polymethacrylates from Designed Bulky Monomers by Anionic and Free Radical Catalyses

Yoshio Okamoto[1], Kumiko Tsunematsu[1], Kiyoko Ueda[1], Yasuaki Hidaka[1], Naotaka Kinjo[1], and Tamaki Nakano[2]

[1] Department of Applied Chemistry, Graduate School of Engineering, Nagoya University, Furo-cho, Chikusa-ku, Nagoya 464-8603, Japan
[2] Graduate School of Materials Science, Nara Institute of Science and Technology (NAIST), Takayama-cho 8916-5, Ikoma, Nara 630-0101, Japan

This chapter describes asymmetric anionic and free-radical polymerization of bulky methacrylates leading to highly isotactic, optically active polymers having a helical conformation with excess screw sense (helix-sense-selective polymerization). The monomers used in this work include 1-phenyldibenzosuberyl methacrylate, racemic and optically active 10,11-*O*-isopropylidene-*trans*-10,11-dihydroxy-5-phenyl-10,11-dihydro-*5H*-dibenzo[*a,d*]cyclo-heptene-5-yl methacrylate, (1-methylpiperidin-4-yl)diphenylmethyl methacrylate, and 10,10-dibutyl-9-phenylanthracen-9-yl methacrylate. These monomers gave highly isotactic, helical polymers not only by anionic mechanism but also by radical mechanism. Helix-sense selection was realized on the basis of chirality of monomer or initiator or additive. Enantiomer selection was also attained in the polymerization of the chiral monomer.

Helical synthetic polymers, especially optically active ones with single screw sense, have been drawing attention not only because the synthesis and the structure are interesting from an academic view but also because this class of polymers has a wide variety of potential application including chiral recognition and liquid crystal formation.[1-3] Helical polymers having excess screw sense are produced from monomers including acrylic compounds, isocyanides, isocyanates, chloral, and acetylene derivatives. We have been extensively studying helix-sense-selective polymerization of methacrylates and related monomers.[1,2] Bulky methacrylates such as triphenylmethyl methacrylate (**TrMA**)[4] afford highly isotactic, optically active polymers having a single-handed helical conformation of the main chain by asymmetric (helix-sense-selective) anionic polymerization.[1-5] The anionic polymerization is carried out using a complex of an organolithium and a chiral ligand such as (-)-sparteine, (+)-1-(2-pyrrolidinylmethyl)pyrrolidine (**PMP**), and (+)-2,3-dimethoxy-1,4-bis(dimethylamino)butane (**DDB**).[4] The optical activity of the polymers is based mostly on the helical structure; therefore, poly(methyl methacrylate)s (PMMAs) derived from the polymers show very low optical activity based only on the asymmetric centers in the vicinity of the chain terminals.[4,5] The single-handed helical polymethacrylates show chiral recognition ability toward a wide range of racemic compounds and some of them have been successfully commerciallized.[6]

In recent years, we are focusing on the synthesis of helical, optically active polymethacrylates via free-radical mechanism as well as anionic mechanism. Although free-radical reaction is generally much more difficult to control compared with anionic reaction, it has a major advantage over anionic reaction that severe dry conditions are usually not necessary and a much wider variety of monomers can be polymerized.[7] This means that single-handed helical polymethacrylate bearing functional groups may be synthesized under relatively mild reaction conditions via radical mechanism if the reaction and monomer structure are adequately designed. Through our efforts towards this goal, helix-sense selection has been realized in polymerization of several monomers on the basis of monomer design or using chiral additive. In this chapter, we discuss the polymerization of four monomers.

Helix-Sense-Selective Radical Polymerization of 1-Phenyldibenzosuberyl Methacrylate (PDBSMA)

PDBSMA gives a highly isotactic, optically active polymer ($[\alpha]_{365}$ +1778°) with completely single-handed helical structure by anionic polymerization using the complexes of *N,N'*-diphenylethylene monolithium amide (**DPEDA-Li**) with **Sp**, **DDB**, and **PMP** in toluene at -78°C.[8] This monomer affords a highly isotactic, helical polymer also by free-radical polymerization regardless of reaction condition.[9] The free-radical polymerization product is considered to be an equimolar mixture of right- and left-handed helices.   The high

PDBSMA

stereospecificity in the **PDBSMA** polymerization by radical catalysis may arise from the rigid structure of the 1-phenyldibenzosuberyl moiety with the two phenyl groups bridged with an ethylene group leading to a rigid helical conformation of growing radical. 1-(2-Pyridyl)- and 1-(3-pyridyl)dibenzosuberyl methacrylates having the same backbone as that of **PDBSMA** also afford highly isotactic polymers by radical polymerization.[8,10]

The highly isotactic structure of the radical polymerization product indicates that introduction of chirality into polymerization reaction may result in helix-sense selection by radical mechanism. Based on this idea, production of either helix in excess was first achieved using an optically active initiator, chain-transfer agent, and solvent (additive) leading to a ratio of enantiomeric helix of up to 7:3.[11]

We recently found that helix-sense-selective radical polymerization of PDBSMA is possible with higher efficiency using an optically active cobalt (II) complex (**1**).[12] Since the complex **1** is a d^7 species (radical), it can interact with growing radical in the polymerization system. Selected conditions and results

1

of polymerization in a CHCl$_3$-pyridine mixture are shown in Table I. The polymerization products were fractionated into tetrahydrofuran (THF)-insoluble (higher-molecular-weight) and -soluble (lower-molecular-weight) parts. The polymerization in the presence of **1** resulted in optically active polymers (THF-soluble part) though the presence of **1** reduced polymer yield and molecular weight of the products. No polymer was obtained when [Co]$_0$ was 0.13 M or in the absence of pyridine. The optically active polymers showed circular dichroism (CD) absorptions with a spectral pattern similar to that of a completely single-handed poly(**PDBSMA**) prepared by the anionic polymerization, indicating that the optical activity of the radically obtained polymers is based on excess screw sense of the main chain helix. SEC analysis of a dextrorotatory polymer using polarimetric and UV detectors provided important information on the chiral structure of the polymer (Figure 1), that is, the higher-molecular-weight fractions showed higher optical activity. This indicates that helical sense excess is larger for higher-molecular-weight fractions. Chromatographic fractionation of the high-molecular-weight part (shaded area in Figure 1A) gave 8wt% of the THF-soluble polymer with a DP of 43 whose polarimetric and UV chromatograms had a very similar shape and the same peak position, indicating that optical activity does not depend on molecular weight in this range (Figure 1B). Specific rotation of this fraction was estimated to be $[\alpha]_{365}$ +1600° and in addition, CD spectral pattern and intensity of this

Table I. Helix-Sence-Selective Radical Polymerization of PDBSMA with AIBN in the Presence of 1 in a Chloroform-Pyridine Mixture at 60°C[a]

Run	$[1]_0$ (M)	Yield[b] (%)	DP[c]	Mw/Mn[c]	THF-soluble part Yield (%)	DP[d]	Mw/Mn[d]	$[\alpha]_{365}^e$ (deg)
1	0	86	170	2.78	3	19	1.27	
2	0.011	59			2	19	1.20	+270
3	0.039	39	78	1.60	3	19[f]	1.28	+550
4	0.057	16			4	19	1.19	+160

[a]Conditions: **PDBSMA** = 0.5 g, $[\text{PDBSMA}]_0$ = 0.44-0.45 M, $[\text{AIBN}]_0$ = 0.029-0.031 M, $[\text{pyridine}]_0$ = 0.51~0.54 M, time = 24 hr. [b]MeOH-insoluble part of the products. [c]Determined by GPC of PMMA derived from poly(**PDBSMA**). [d]Determined by GPC of poly(**PDBSMA**). [e]Estimated based on GPC curves obtained by UV and polarimetric detections. [f]DP = 20 (Mw/Mn = 1.14) as determined by GPC of PMMA.

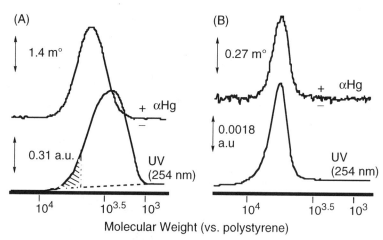

Figure 1. GPC curves obtained by polarimetric (top) and UV (bottom) detectors of the THF-soluble polymer of run 3 in **Table I** (A) and the high-molecular-weight fraction separated from the THF-soluble polymer (B).

fraction were quite similar to those of the anionically obtained single-handed helical polymer. This means that the polymer of this fraction has almost single-handed conformation.

The results described so far indicate that the Co complex **1** induced a single-handed helical conformation in the radical polymerization of **PDBSMA**. This induction is probably based on the interaction between **1** and growing polymer radical. Right- and left-handed helical radicals may have different strength of interaction or binding constant with **1**, that is, **1** may undergo stronger interaction with one helix than with the other, leading to different apparent propagation rates of the two radicals (Figure 2). This assumption will result in dependence of optical activity on molecular weight.

The complex **1** was recently shown to induce configurational chirality of the main-chain in the polymerization of *N*-phenyl and *N*-cyclohexyl maleimides.[13]

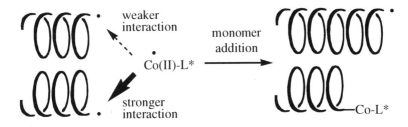

Figure 2. Proposed mechanism of helix-sense-selective radical polymerization of **PDBSMA** using an optically active Co(II) complex.

Polymerization of a Chiral PDBSMA Analogue

(±)- and (+)-10,11-*O*-Isopropylidene-*trans*-10,11-dihydroxy-5-phenyl-10,11-dihydro-5*H*-dibenzo[*a,d*]cycloheptene-5-yl methacrylate (**IDPDMA**), a chiral **PDBSMA** derivative, was synthesized and polymerized.[14] This novel monomer was designed to undergo isotactic-specific radical polymerization similar to that of **PDBSMA** and the monomer in optically active form was expected to lead to helix-sense selection in the polymerization. **IDPDMA** was prepared in racemic form by four-step reactions starting from dibenzosuberenone, and optically active monomer was obtained by HPLC resolution ((+)-isomer: e.e. ~100%; $[\alpha]_{365}$ +548°).

IDPDMA

Table II. Anionic Polymerization of (±)- and (+)-IDPDMA Using DPEDA-Li Complexes at -78°C[a]

Run	Monomer	Solvent	Ligand	Time (hr)	Polymer Yield[b] (%)	$[\alpha]_{365}$[c]	DP[d]	E.e. (%)[e] (monomeric units)	Recovered monomer E.e. (%)[f]
1	racemic	Toluene	(-)-Sp	24	27	n.d.[g]	159	+13.5	-5.1
2	racemic	Toluene	(+)-PMP	24	20	+655°	30	+21.4	-5.3
3	racemic	Toluene	(+)-DDB	4	11	+866°	20	+67.0	-8.0
4	racemic	Toluene	(+)-DDB	8	22	+973°	29	+44.0	-12.4
5	(+)-	THF	none	24	54	+1540°[i]	78		

[a][M]/[I] = 20. [b]MeOH-insoluble part. [c]In CHCl₃ (conc. 0.5 g/dl). [d]Determined by GPC analysis of PMMAs derived from the original polymers. [e]Calculated from e.e. of the monomer in feed, polymer yield, and e.e. of the remaining monomer. [f]Determined by HPLC resolution. [g]Products were insoluble in THF or CHCl₃. [i]$[\alpha]_D^{25}$ +353°.

SOURCE: Reproduced from Reference 14. Copyright 1999 The Society of Polymer Science, Japan.

The conditions and results of anionic polymerization of (±)- and (+)-**IDPDMA** using **DPEDA-Li** complexes of (-)-**Sp**, (+)-**DDB**, and (+)-**PMP** are shown in Table II. The anionic polymerization was slower compared with the polymerization of **PDBSMA** under similar reaction conditions probably because of the additional bulkiness of **IDPDMA** compared with **PDBSMA**. The polymerizations of racemic monomer using optically active initiators resulted in polymers with highly isotactic configuration (mm > 99%) and high optical activity. The values of specific rotation were larger than expected from the e.e. of the monomeric unit; hence the polymers are considered to have excess helicity.

DPEDA-Li   (-)-Sp   (+)-DDB   (+)-PMP

By polymerization without a chiral ligand (run 5), the optically pure (+)-**IDPDMA** gave a polymer whose optical activity was comparable to that of the single-handed helical poly(**TrMA**) and much larger than that of the starting monomer, suggesting that this polymer may have a single-handed helicity induced by the chirality of the monomer during polymerization. This is supported by the CD spectrum of the polymer (Figure 3). The spectral pattern was different from that of the monomer and overall similar to those of single-handed helical poly(**PDBSMA**)[8] and poly(**TrMA**).[4b] The resemblance was especially obvious in the wavelength range below 260 nm with an intense positive peak at around 240 nm. Some minor differences may be due to the contribution by the (+)-monomeric unit.

In the polymerization of racemic monomer, the unreacted monomer was found to be optically active, indicating that the polymerization was enantiomer selective. (-)-**Sp**, (+)-**DDB**, and (+)-**PMP** preferentially polymerized (+)-monomer over (-)-monomer. **DDB** among the three chiral ligands led to the highest selectivity. It has been reported that helical conformation of the growing species has an important role in enantiomer selection in the asymmetric anionic polymerization of phenyl-2-pyridyl-o-tolylmethyl methacrylate (**PPyoTMA**).[15] In the **PPyoTMA** and **IDPDMA** polymerizations under similar conditions using **DDB** as chiral ligand, e.e. of the unreacted monomer was 8.5% and 12.4%, respectively, at 22% yield, indicating that enantiomer selectivity was higher in the **IDPDMA** polymerization. This means that the growing anion in the **IDPDMA** polymerization system may have a more rigid and complete helical structure than the poly(**PPyoTMA**) anions having a lower isotacticity (mm 94%), which may have a connection with the higher selectivity.

Free-radical polymerization of **IDPDMA** having various e.e.'s was carried out using AIBN as initiator in toluene (Table III). Regardless of the e.e. of the starting monomer, nearly completely isotactic polymers were obtained. The polymers are considered to have a helical conformation. The THF-soluble part

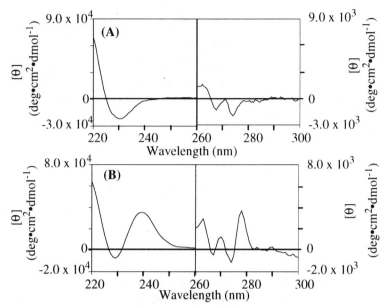

Figure 3. CD spectrum of (+)-**IDPDMA** with 100% e.e.
(A) and that of poly([(+)-**IDPDMA**]) obtained in run 5 in
Table II (B) [THF, r.t.].
SOURCE: Reproduced from Reference 14. Copyright 1999
The Society of Polymer Science, Japan.

Table III. Radical Polymerization of IDPDMA Using AIBN in Toluene at 60°C[a]

| | | Polymer | | | | Recovered monomer |
Run	E.e. (%) of monomer in feed[b]	Yield[c] (%)	DP[d]	Tacticity(%)[e] mm	E.e. (%) of monomeric units[f]	E.e. (%)[b]
1	0	75	202	>99	0	0
2	−23.8	63	106	>99	−28.9	−14.9
3	−48.0	63	162	>99	−58.9	−29.8
4	−64.7	59	128	>99	−78.0	−46.0
5	−74.0	61	121	>99	−82.3	−61.1
6	+100	28	217	>99	+100	+100

[a][M]/[I] = 50, monomer 0.1 g, toluene 2.3 ml, time 24 hr. [b]Determined by HPLC resolution. [c]Hexane-insoluble part. [d]Determined by GPC analysis of PMMA derived from the original polymer. [e]Determined by 400 MHz ^{1}H NMR analysis of PMMAs derived from the original polymers. [f]Calculated from e.e. of the monomer in feed, polymer yield, and e.e. of the remaining monomer.
SOURCE: Reproduced from Reference 14. Copyright 1999 The Society of Polymer Science, Japan.

(DP = 37, Mw/Mn = 1.09) of the polymerization products of optically pure (+)-**IDPDMA** (run 6) showed CD absorptions whose spectral shape seemed to be intermediate between those of (+)-**IDPDMA** monomer and the poly[(+)-**IDPDMA**] prepared by the anionic polymerization. This suggests that the chiroptical properties of the THF-soluble poly[(+)-**IDPDMA**]s are based on both the chiral structure of the ester group and the main-chain helical conformation with excess single-handed helicity which was induced by the effect of the chiral ester group during the radical polymerization process. However, the helix-sense excess of the poly[(+)-**IDPDMA**] prepared by radical polymerization may be lower than that of the polymer obtained by anionic polymerization because the peak intensity in the CD spectrum around 240 nm for the radically obtained polymer (Figure 4) is much smaller than that of the anionically obtained polymer (Figure 3 B).

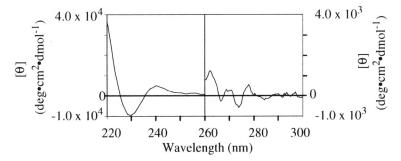

Figure 4. CD spectrum of THF-soluble poly[(+)-**IDPDMA**] obtained in run 6 in Table III [THF, r.t.].

SOURCE: Reproduced from Reference 14. Copyright 1999
The Society of Polymer Science, Japan.

Enantiomer selection was also found in the radical polymerization (Figure 5). The e.e. of the recovered monomer was lower and that of the monomeric unit in polymer was higher than that of the starting monomer at all optically purity of the feed monomer, indicating that the growing radical preferentially reacted with the enantiomer, that is, enantiomer-selective polymerization took place. The enantiomer selection may be governed by the excess helicity of the growing radical. The growing species consisting of excess enantiomeric component of monomeric units probably takes helical conformation with excess helix sense which can choose one enantiomer of **IDPDMA** over the other.

Enantiomer selection by radical mechanism has been attained in the polymerization of PPyoTMA having various e.e.'s.[16] The selectivity observed in this work was higher compared with that in the radical polymerization of **PPyoTMA** (-53.4%e.e.) where the recovered monomer at 95.5% polymer yield had an optical purity rather close to that of the starting monomer (-50.3%e.e.). In contrast, in the polymerization of **IDPDMA** (-48.0%e.e.), the e.e. of the recovered monomer was -29.8% even at a lower polymer yield (63%). The higher selectivity may be ascribed to a higher conformational regularity of the

growing species in the **IDPDMA** polymerization due to the highly isotactic structure.
  The enantiomer selectivity observed in the radical polymerization was much lower than those observed for other systems based on anionic catlysis.[17] The absence of coordination mechanism in free radical polymerization is considered to be responsible for the rather low selectivity.

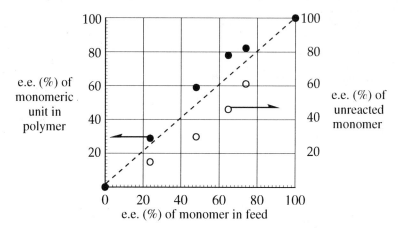

Figure 5. Relation between e.e. (absolute value) of monomeric unit in polymer (●) and unreacted monomer (○) and that of monomer in feed corresponding to the data shown in **Table III**.
SOURCE: Reproduced from Reference 14. Copyright 1999 The Society of Polymer Science, Japan.

Design of Monomers for Highly Isotactic Specific Radical Polymerization

  The isotactic specificity in the radical polymerization of bulky methacrylates depends strongly on monomer structure.[7] As described so far, **PDBSMA** backbone has been shown to realize a highly isotactic specific propagation. The high stereospecificity has been ascribed to the rigid structure of the 1-phenyldibenzosuberyl moiety. We recently found that two new monomers that do not have dibenzosubery structure in the side chain, (1-methylpiperidin-4-yl)diphenylmethyl methacrylate (**MP4DMA**)[18] and 10,10-dibutyl-9-phenyl anthracen-9-yl methacrylate (**DBPAMA**),[18b,19] give highly isotactic polymers by the radical polymerization as well as by anionic polymerization.
  Asymmetric anionic polymerization of **MP4DMA** using **DPEDA-Li-PMP** and **DPEDA-Li-DDB** in toluene at -78°C produced highly isotactic, optically active polymers ($[\alpha]_{365}$ +977° (**PMP** system); +1053° (**DDB** system)). These polymers are considered to have helical conformation with excess screw sense. This monomer also gave a highly isotactic polymers by radical polymerization (mm = 98% (reaction at 60°C); 95% (reaction at 0°C)). The radical polymerization products are reasonably assumed to be an equimolar mixture of right- and left-handed helices. This high stereospecificity is quite in contrast to the moderate isospecificity observed in the radical polymerization of

cyclohexyldiphenylmethyl methacrylate (**CHDPMA**) (mm = 45% in polymerization at 40°C).[20] The methyl group attached to the N atom in **MPD4MA** may play a crucial role in the stereoregulation though the exact mechanism is not clear yet.

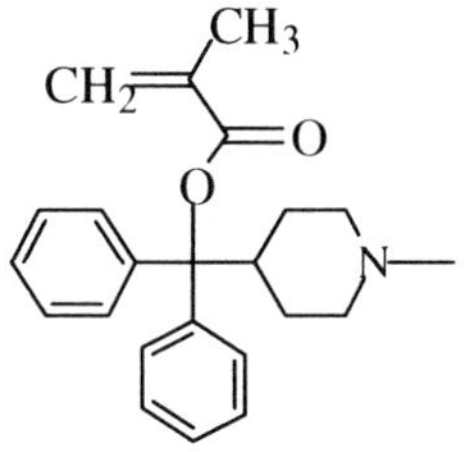

MPD4MA 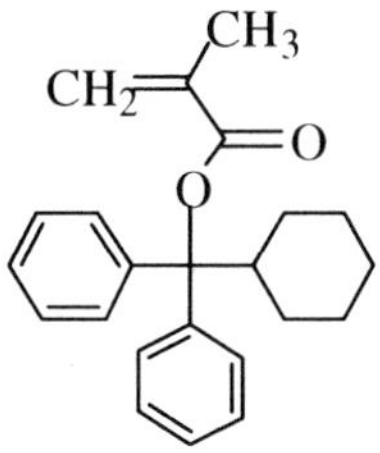   CHDPMA

The radical polymerization of **MP4DMA** was helix-sense selective when the reaction was carried out in the presence of (-)-menthol. The polymerizations in toluene using AIBN at 60°C in the presence of (-)-menthol at [menthol]$_o$/[monomer]$_o$ = 5 and 1 gave optically active, highly isotactic polymers showing $[\alpha]_{365}$ +87° and +3°, respectively. The optical activity of the polymer increased with a decrease in polymerization temperature: at 0°C at [menthol]$_o$/[monomer]$_o$ = 5, specific rotation of the polymer reached $[\alpha]_{365}$ +106°. The optical activity was considered to be based on excess screw sense of the main chain induced during the polymerization because the CD spectrum of the radically obtained polymer showing $[\alpha]_{365}$ +87° had a similar pattern to that of the anionically obtained polymer showing $[\alpha]_{365}$ +977°. Although the mechanism of the helix-sense selection in this case is not clear, such a selection using menthol has been already performed in the radical polymerization of **PDBSMA** through helix-sense-selective termination (hydrogen radical transfer from menthol to a growing radical).[11]

DBPAMA also afforded an optically active, highly isotactic polymer ($[\alpha]_{365}$ +183°, mm > 98%) by asymmetric anionic polymerization in toluene at -40°C using **DPEDA-Li-(+)-DDB** complex as an initiator. SEC and chromatographic resolution analyses of the polymer indicated that the optical activity was not clearly dependent on molecular weight and the (+)-polymer

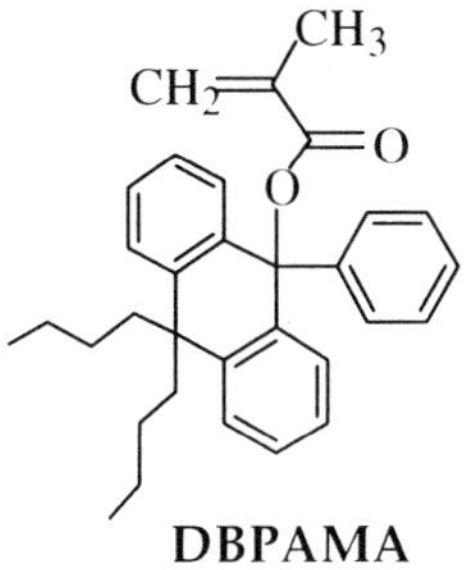

DBPAMA

sample contained no levorotatory fraction. A single-handed helical structure is strongly suggested by these observations. However, the specific rotation of the polymer was much smaller than that of single-handed helical poly(**TrMA**).[4] This may mean that a side-chain propeller conformation of three aryl groups, which has been suggested for poly(**TrMA**)[6] but is impossible for poly(**DBPAMA**), contributes to specific rotation of poly(**TrMA**).

Radical polymerization of **DBPAMA** led to highly isotactic polymers. The isotacticity increased with a decrease in temperature and the polymers obtained at 0°C and -20°C were highly isotactic (mm > 98%, DP = 974 at 0°C; mm ~99%, DP = 563 at -20°C). It is interesting that the isotactic, helical poly(**DBPAMA**)s obtained by radical polymerization having relatively high DP (up to 974) were soluble in $CHCl_3$ and tetrahydrofuran. This is in contrast to the fact that helical polymethacrylates with high DP generally have tendency to form aggregates and become hardly soluble.[4a,21] The side-chain butyl chains of poly(**DBPAMA**) may prevent intermolecular interaction. The good solubility of poly(**DBPAMA**) would make it possible to clarify solution properties of high-molecular-weight, helical vinyl polymers.

Helix-sense selection was realized in the radical polymerization by the use of optically active neomenthanethiol as an optically active chain-transfer agent and an optically active polymer ($[\alpha]_{365}$ +74°) was obtained.

Conclusions

   PDBSMA afforded an isotactic, single-handed helical polymer not only by anionic polymerization using complexes of **DPEDA-Li** with chiral ligands but also by radical polymerization using Co complex **1**. Asymmetric anionic polymerization of **IDPDMA**, **MP4DMA**, and **DBPAMA** led highly isotactic, helical polymers having excess helicity. Radical polymerization of these three monomers was also found to be highly isotactic-specific. Helix-sense selection in the radical polymerization of optically active **IDPDMA** was achieved on the basis of the chirality of the monomer, that of **MP4DMA** using menthol as a chiral additive, and that of **DBPAMA** using neomenthanethiol as a chain-transfer agent. In the anionic and radical polymerization of **IDPDMA**, enantiomer-selection was observed. Thus, we have shown that by properly designing monomer structure and reaction conditions, helix-sense-selective and enantiomer-selective polymerization were possible via radical mechanism as well as via anionic mechanism.

   A goal of our series of studies in this area is to produce and utilize single-handed helical polymers having a wide variety of functional groups in the side chain. Such polymers would be directly prepared from functional monomers by radical polymerization without using protection/deprotection technique which is necessary in anionic synthesis. To reach the goal, it would be important to establish catalytic stereoregulation methods based on radical chemistry that can be applied generally to a wide range of monomers.

Acknowledgment. This work was supported in part by the New Energy and Industrial Technology Development Organization (NEDO) under the Ministry of International Trade and Industry (MITI), Japan, through the grant for "Precision Catalytic Polymerization" in the Project "Technology for Novel High-Functional Material" (1996-2000) and in part by Grant-in-Aid for

23

Scientific Research (No.11750762) from the Ministry of Education, Science, Sports, and Culture, Japan.

Literature Cited

(1) Okamoto, Y.; Nakano, T. *Chem. Rev.,* **1994**, *94*, 349.
(2) (a) Nakano, T.; Okamoto, Y. In *The Polymeric Materials Encyclopedia;* Salamone, J. C. Ed.; CRC press: Florida, 1996; pp. 417-423. (b) Nakano, T.; Okamoto, Y. In *Catalysis in Precision Polymerization;* Kobayashi, S. Ed.; John Wiley & Sons: Chishester, Sussex, 1997; pp. 271-309.
(3) Wulff, G. *Angew. Chem. Int. Ed. Engl.* **1995**, *34*, 1812.
(4) (a) Okamoto, Y.; Suzuki, K.; Ohta, K.; Hatada, K.; Yuki, H. *J. Am. Chem. Soc.* **1979**, *101*, 4763. (b) Nakano, T.; Okamoto, Y.; Hatada, K. *J. Am. Chem. Soc.* **1992**, *114*, 1318.
(5) Nakano, T.; Okamoto, Y. *Macromol. Chem. Rapid Commun.* **2000**, in press.
(6) (a) Okamoto, Y. *CHEMTECH*, **1987**, 144. (b) Okamoto, Y.; Hatada, K. *J. Liq. Chromatogr.* **1986**, *9*, 369.
(7) (a) Nakano, T.; Okamoto, Y. In *Controlled Radical Polymerization (ACS) Symposium Series 685*; Matyjazsewski, K., Ed.; ACS: Washington D.C., 1998; p 451-462. (b) Hatada, K.; Kitayama, T.; Ute, K. *Prog. Polym. Sci.* **1988**, *13*, 189. (c) Yuki, H.; Hatada, K. *Adv. Polym. Sci.* **1979**, *31*, 1.
(8) Nakano, T.; Matsuda, A.; Mori, M.; Okamoto, Y. *Polym. J.* **1996**, *28*, 330.
(9) (a) Nakano, T.; Mori, M.; Okamoto, Y. *Macromolecules* **1993**, *26*, 867. (b) Nakano, T.; Matsuda, A.; Okamoto, Y. *Polym. J.* **1996**, *28*, 556.
(10) Nakano, T.; Satoh, Y.; Okamoto, Y. *Polym. J.* **1998**, *30*, 635.
(11) (a) Nakano, T.; Shikisai, Y.; Okamoto, Y. *Polym. J.* **1996**, *28*, 51. (b) Nakano, T.; Shikisai, Y.; Okamoto, Y., *Proc. Japan Acad.* **1995**, *71*, Ser. B , 251.
(12) Nakano, T.; Okamoto, Y. *Macromolecules* **1999**, *32*, 2391.
(13) Nakano, T.; Tamada, D.; Miyazaki, J.-i.; Kakiuchi, K.; Okamoto, Y. *Macromolecules* **2000**, *33*, 1489.
(14) Nakano, T.; Kinjo, N.; Hidaka, Y.; Okamoto, Y. *Polym. J.* **1999**, *31*, 464.
(15) Yashima, E.; Okamoto, Y.; Hatada, K. *Macromolecules* **1988**, *21*, 854.
(16) Okamoto, Y.; Nishikawa, M.; Nakano, T.; Yashima, E.; Hatada, K. *Macromolecules* **1995**, *28*, 5135.
(17) For a review, see: Okamoto, Y.; Yashima, E. *Prog. Polym. Sci.* **1990**, *15*, 263.
(18) (a) Ueda, K.; Nakano, T.; Okamoto, Y. *Polym. Prepr. Jpn.* **1999**, *48(7)*, 1327. (b) Nakano, T.; Tsunematsu, K.; Ueda, K.; Kinjo, N.; Okamoto, Y. The 5th International Symposium on Polymers for Advanced Technologies (Tokyo, 1999) Preprint, 291.
(19) (a) Kinjo, N.; Nakano, T.; Okamoto, Y. . *Polym. Prepr. Jpn.* **1999**, *48(2)*, 155. (b) Nakano, T.; Tsunematsu, K.; Kinjo, N.; Okamoto, Y. IUPAC International Symposium on Ionic Polymerization (Kyoto, 1999) Preprint, 67.
(20) Okamoto, Y.; Nakano, T.; Fukuoka, T.; Hatada, K. *Polym. Bull.* **1991**, *26*, 254.

Chapter 3

Profile of the Helical Structure of Poly(propiolic esters)

Ryoji Nomura, Hideo Nakako, Yasunori Fukushima, and Toshio Masuda

Department of Polymer Chemistry, Graduate School of Engineering, Kyoto University, Kyoto 606–8501, Japan

Poly(propiolic esters) with chiral substituents have proven to exist in well-ordered helical conformation with an excess of one-handed screw-sense. The main chain of poly(propiolic esters) consists of relatively large helical domains interposed between disordered states, which was evidenced by the chiroptical properties of copolymers from chiral and achiral comonomers. NMR study of various poly(propiolic esters) enabled estimation of not only the activation energy of helix reversal but also the free energy difference between the helical and disordered states.

Much attention has been paid to the synthesis of helical polymers owing to their unique functions such as molecular recognition ability and catalytic activity for asymmetric syntheses (1–6). Recent progress in polymer synthesis has enabled it to afford a variety of well-ordered helical polymers with an excess of one-handed screw-sense. Among them, helical polymers with π-conjugation along the main chain are very interesting since they are potentially useful as polarization-sensitive electrooptical materials, asymmetric electrodes, and so forth (7).

In most cases, the helicity of π-conjugated polymers does not originate from the twist of the main chain, and the aggregated forms of plural polymer chains possess a helical structure that contributes to their large chiroptical properties. An exceptional example is given by substituted polyacetylenes whose main chains can adopt helical conformation even if they do not aggregate. Helical conformation with an excess of one-handed screw-sense is attained by incorporating appropriate chiral substituents onto highly stereoregular cis polymers from monosubstituted acetylenes (8–14). However, in contrast to the stiffness of polyisocyanates that possess a very large helical domain (15), the main chain of substituted polyacetylenes is rather flexible, and thus, they possess a helical domain of very short persistence length. The flexibility of the main chain results in the observation of clear CD effects only at low temperature for poly(1-alkynes) with chiral substituents (8). Another proof of the short persistence length of substituted polyacetylenes is given by the experimental and computational study on the conformation of poly(phenylacetylenes) demonstrated by Yashima et al (13). Therefore, polyacetylenes cannot maintain a well-defined helical structure in solution unless they possess bulky substituents.

Apart from these profiles of helical poly(1-alkynes) and poly(phenylacetylenes), our recent conformational study of poly(propiolic esters), another type of substituted polyacetylenes, led to an unexpected conclusion that they possess a large helical domain size. In other words, poly(propiolic esters) exist in helical conformation in solution even if they do not have bulky substituents. Furthermore, we found that thermodynamic and kinetic parameters, governing the stability of the helix, are readily obtained by a simple NMR technique (16–18). In the present review article, we wish to briefly summarize our recent study on the secondary structure of poly(propiolic esters).

Approach to Well-Ordered Helical Poly(propiolic esters)

Monomer Design

The most significant factor that influences the secondary conformation of poly(propiolic esters) is the length of the alkylene spacer between the ester group and the chiral center (17). This effect is clearly recognized by comparison of the chiroptical properties of poly(1)–poly(5) that possess an identical stereogenic group but a different number of alkylene spacers (Scheme 1). For example, polymers with alkylene spacers, poly(2)–poly(5), prepared with Rh catalyst, display very large optical rotations ([α]$_D$), while the [α]$_D$ of the polymer without any alkylene spacer, poly(1), is very small (Table I). In a similar way, very intense CD signals are detected for poly(2)–poly(5), which is

Scheme 1

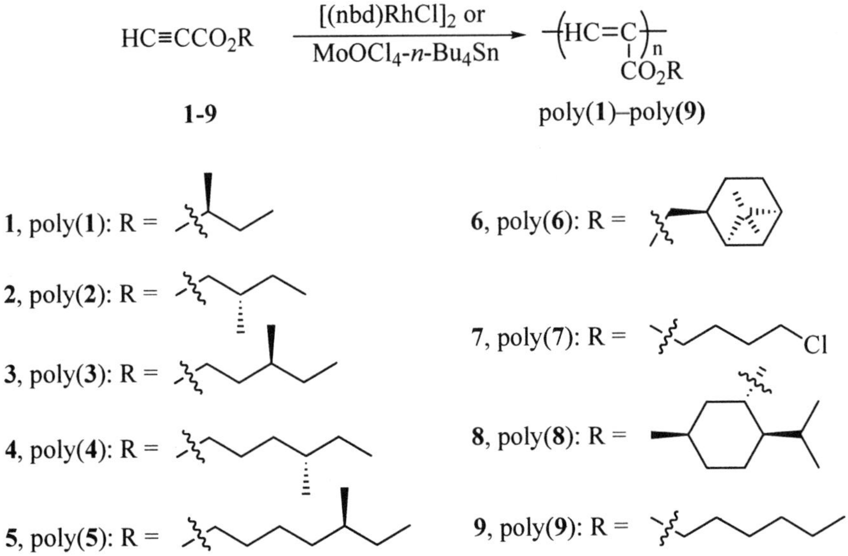

$$HC \equiv CCO_2R \xrightarrow[\text{MoOCl}_4\text{-}n\text{-Bu}_4\text{Sn}]{\text{[(nbd)RhCl]}_2 \text{ or}} \left(HC = \underset{\underset{CO_2R}{|}}{C} \right)_n$$

1-9   poly(1)–poly(9)

1, poly(1): R =

2, poly(2): R =

3, poly(3): R =

4, poly(4): R =

5, poly(5): R =

6, poly(6): R =

7, poly(7): R =

8, poly(8): R =

9, poly(9): R =

in contrast to the very weak Cotton effect of the CD spectrum of poly(1) (Figure 1a). The main chain of poly(2)–poly(5), therefore, exists in helical conformation, and the domain size of one-handed helical conformation predominates over that of the counterpart. Emphasis should be placed on the difference in the CD pattern of poly(1) from those of poly(2)–poly(5): all the polymers with alkylene spacers, poly(2)–poly(5), show Cotton effects with different intensity but with identical patterns, whereas the CD effect of poly(1) significantly differs in shape from those of poly(2)–poly(5) as shown in Figure 1a. Therefore, the poorer chiroptical property of poly(1) is not simply due to the ill-defined screw-sense. As shown in Figure 1b, one can recognize the red-shifted UV-visible absorption of poly(1) compared with the spectra of the other polymers. Since the cutoff wavelength directly relates to band gap energy, *i.e.*, the degree of main-chain conjugation, the red shift observed in poly(1) is attributed to its extended main-chain conjugation. In other words, the coplanarity of the polymer backbone is enhanced when the alkylene spacer is eliminated. The main chain of polymers with alkylene spacers is, thus, more tightly twisted from the planar structure, while, as discussed below, poly(1) possesses an ill-ordered secondary conformation. This difference in secondary structure causes a difference in the shape of the CD signal of the polymers. Therefore, an alkylene spacer between the ester group and the stereogenic carbon is necessary for construction of a well-ordered helical structure.

Table I. Polymerization of Various Chiral Propiolic Esters with [(nbd)RhCl]$_2$^{a}

Monomer	$[\alpha]_D^b$ (°)	Yield^c (%)	$M_n/10^{3\,d}$	$[\alpha]_D^e$ (°)
1	+18	36	21	+4
2	+5	36	80	−473
3	+11	37	24	−612
4	+12	30	76	−418
5	+6	30	60	−358
6	−28	15	99	−340

^{a} Conditions; [[(nbd)RhCl]$_2$] = 20 mM [M]$_0$ = 1.0 M, acetonitrile, 30 °C, 24 h. ^{b} In CHCl$_3$, c = 0.2 g/dL. ^{c} Methanol-insoluble part. ^{d} Estimated by GPC (PSt, CHCl$_3$). ^{e} In CHCl$_3$, c = 0.06 g/dL.

As demonstrated in a previous conformational study of poly(phenylacetylenes), introduction of bulky substituents enhances the persistence length of the helical domain (*13*). Such a tendency is also recognized in poly(propiolic esters) (*17*). Comparison of the CD spectra of two polymers, poly(6) and poly(2), that have an alkylene spacer of identical length but different bulkiness clearly exemplifies this effect (Figure 2). Thus, the

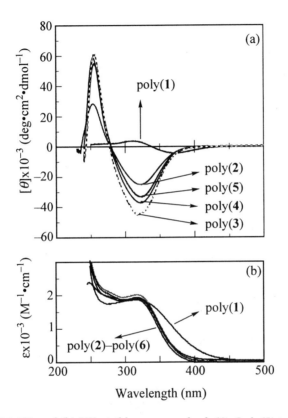

Figure 1. *(a) CD and (b) UV-visible spectra of poly(1)_Poly(5) in CHCl₃ (c =*
6.0x10⁻⁴ mol/L, 20 °C).
(Reproduced with permission from reference 17. Copyright 2000 American
Chemical Society.)

introduction of bulky myrtanyl group increases the absolute magnitude of the Cotton effect.   No difference in the CD signal pattern between poly(2) and poly(6) shows that the bulkiness of the pendant does not influence the secondary conformation of polymer.

Catalyst

Two types of catalyst are effective for the polymerization of propiolic esters. One is the Rh catalyst (19), [(nbd)RhCl]₂, and the other involves Mo- and W-based metathesis catalysts (20).   The ability of the Rh catalyst to provide polymers with excellent cis-transoidal stereoregularity allows the formation of well-ordered helical poly(propiolic esters) (17).   On the other hand, polymers prepared with the Mo catalyst showed no signal due to cis olefinic protons in their ^{1}H NMR spectra, indicating the lack of stereoregularity.   This poorly controlled first-order structure results in the deflated chiroptical properties of the Mo-based polymers (Table II); no distinct CD effect is detected for the polymers with alkylene spacers, poly(2)_poly(5).

Stability of the Helical Structure of Poly(propiolic esters)

Sergeant and Soldiers Principle and Majority Rule

From the results mentioned above, a general methodology for the design of well-ordered helical poly(propiolic esters) is now available.   The next question lies on the stability of the helix, *i.e.*, the persistence length of the helical domain of the polymers.   The large helical domain size of poly(propiolic esters) compared with previously prepared chiral substituted polyacetylenes was qualitatively demonstrated by chiral amplification phenomena in the copolymerization of chiral with achiral comonomers (18), the sergeant-and-soldiers principle (21).   A good example is given by the copolymerization of (–)-myrtanyl propiolate (6) with 4-chlorobutyl propiolate (7). As illustrated in Figure 3, the presence of 3_10% of a chiral segment in the copolymer effectively leads to a large optical rotation of the copolymer that is comparable to that of a homopolymer from 6. In a similar way, even a copolymer with the ratio of 7 : 6 = 89 : 11 shows very intense CD signals whose molar ellipticity is almost the same as that of poly(6).   These results are in contrast to the

Table II. Polymerization of Various Chiral Propiolic Esters with MoOCl$_4$–n-Bu$_4$Sn[a]

Monomer	Yield (%) [b]	$M_n/10^{3}$ [c]	$[\alpha]_D$ [d]
1	43	130	–19
2	56	12	–9
3	52	18	0
4	68	5.1	–4
5	52	12	0
6	84	8.1	–16

[a] Conditions; [MoOCl$_4$] = [n-Bu$_4$Sn] = 20 mM, [M]$_0$ = 0.50 M, in toluene, 60 °C, 24 h. [b] Methanol-insoluble part. [c] Estimated by GPC (PSt, CHCl$_3$). [d] In CHCl$_3$ (c = 0.06 g/dL).

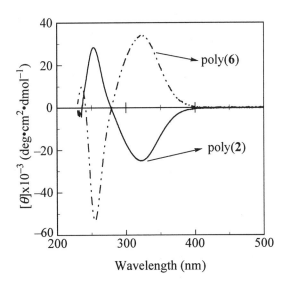

Figure 2. CD spectra of poly(**2**) and poly(**6**) (in CHCl$_3$, c = 6.0x10^{-4} mol/L, 20 °C).

(Reproduced with permission from reference 17. Copyright 2000 American Chemical Society.)

copolymers from phenylacetylenes: no distinct CD signal is attainable from a copolymer containing 10% of a chiral comonomer, and bulky ring substituents are required to effectively achieve the sergeants-and-soldiers principle for the poly(phenylacetylene)-based copolymers (13). A similar chiral amplification is observed in the polymerization of 2 with various optical purities (Figure 4). Namely, there is a positive nonlinear relationship between the enantiomeric excess of monomers and observed optical rotations as well as ellipticity of the Cotton effects of the polymer (the majority rule). Poly(propiolic esters), thus, possess a larger helical domain size than poly(phenylacetylenes) do.

As mentioned above, there exists a distinct difference in the secondary structure of the polymers with and without an alkylene spacer. This structural difference causes very unique behavior of the copolymerization of (–)-menthyl propiolate (8) with hexyl propiolate (9). As illustrated in Figure 5, the copolymer of 8 with 9 shows a nonlinear effect of the chiroptical properties at low chiral contents. However, when the chiral comonomer content exceeds 20%, the increase in the chiral unit leads to a decrease in the optical rotation of the copolymer, and almost no chirality is recognized when the ratio of 8/9 is 60/40 (+10.9° in CHCl$_3$, c = 0.066 g/dL). The $[\alpha]_D$ again increases as the content of the chiral unit is further increased. A similar phenomenon is also observed in the CD spectra of the copolymers (Figure 6). The intensity of CD decreases with an increase in the chiral segment (Figure 6a, b), and no CD effect was obtained when 60% of 8 is incorporated to the copolymer (Figure 6c). The magnitude of the CD signal increases again as the content of 8 increases (Figure 6d, e). Simultaneously, the shape of CD remarkably changes, and the peak top of the CD band shifts toward a long wavelength region. The shape of the CD spectra of the copolymers with low chiral component (Figures 6a, b) is completely identical or just mirror-imaged to those of homopolymers from chiral monomers such as 2–5. On the other hand, the chiral segment-rich copolymers showed CD signals that were very close in shape to that of poly(8) (Figures 6e, d).

The poor chiroptical property of the 8/9 = 60/40 copolymer is not due to lack of stereoregularity because this copolymer is confirmed to possess perfect stereoregularity (cis) by the ^{1}H NMR spectrum in CD$_2$Cl$_2$. A hypothesis that the opposite preferences of 8 and 9 in the screw-sense give the same amount of both helix senses is also ruled out because there is no difference in the sign of optical rotation between 8-rich and 9-rich copolymers. Thus, the poor chiroptical property of the copolymer originates from its randomly coiled, disordered conformation, which results from the difference in the secondary structure between these homopolymers. This conclusion is also supported by the molecular dependence of $[\eta]$ of these copolymers. Namely, the slope of the Mark-Hauwink-Sakurada plot gradually decreases from 1.233 to 0.534 when the chiral content is increased from 0 to 60%, and then increases again to 0.60 with an increase in the segment of 8. This clearly means that poly(9) possesses a

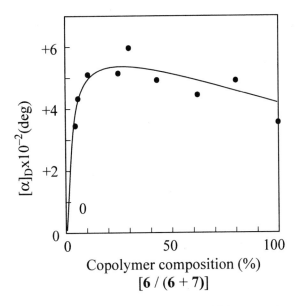

*Figure 3.   Plot of the optical rotation (c = 0.07 g/dL) versus the copolymer composition of poly(**6**-co-**7**).*
(Reproduced with permission from reference 18. Copyright 2000 American Chemical Society.)

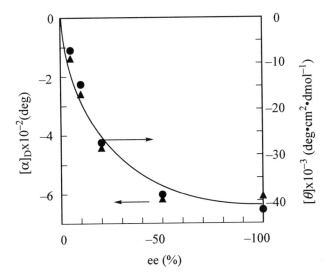

*Figure 4. Plot of the optical rotation (c = 0.07 g/dL) and molar ellipticity at 324 nm (c = 3.8-6.8x10⁻⁴ mol/L) of poly(**2**) versus the ee of **2**.*
(Reproduced with permission from reference 18. Copyright 2000 American Chemical Society.)

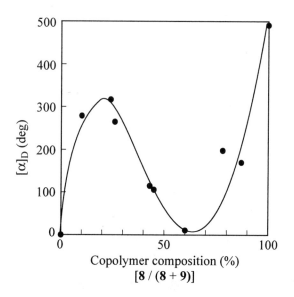

Figure 5. Plots of the optical rotation versus the segment ratio of poly(8-co-9) (in CHCl₃, c = 0.06 g/dL at room temperature).
(Reproduced with permission from reference 18. Copyright 2000 American Chemical Society.)

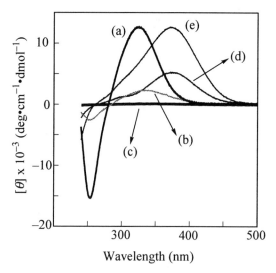

Figure 6. CD Spectra of poly(8-co-9) in CHCl₃ (c = 0.012 g/dL at room temperature). 8/9 = (a) 22/78, (b) 55/45, (c) 60/40, (d) 76/24, (e) 0/100.
(Reproduced with permission from reference 18. Copyright 2000 American Chemical Society.)

stiff main chain due to its well-ordered helical structure and that the copolymer of **8** with **9** (**8**/**9** = 60/40) exists in randomly coiled conformation. The slope of the Mark-Hauwink-Sakurada plot of poly(**8**) was unexpectedly small. However, the very large $[\alpha]_D$ and intense CD effects of poly(**8**) cannot be explained without assuming that poly(**8**) possesses a helical structure. Thus, poly(**8**) exists in helical conformation, but its helical domain size is probably quite small.

NMR Study on the Stability of Helix

The most characteristic feature of poly(propiolic esters) is that the parameters governing the nature of the helix are readily estimated by a simple NMR technique. That is, not only the activation energy for the interconversion process between right- and left-handed helices ($\Delta G^{\ddagger}$), but also the free energy difference between the helical and helix reversal states (ΔG_r) can be determined by NMR spectra.

Activation Energy ($\Delta G^{\ddagger}$) of the helix-helix interconversion

The facile determination of $\Delta G^{\ddagger}$ for various poly(propiolic esters) is based on a result that all poly(propiolic esters) having α-methylene groups show two diastereotopic signals due to the α-methylene protons. This peak separation was reasonably explained by the slow process of the interconversion between right- and left-handed helices on the NMR time-scale at ambient temperature (*19*). The clearest proof is that the α-methylene proton is not diasterotopic when the polymer exists in disordered conformation. Namely, the copolymer (**8**/**9** = 60/40) with random conformation shows α-methylene protons as one broad resonance, while this signal attributed to the α-methylene protons in the well-ordered helical poly(**9**) is clearly separated (Figure 7). The variable temperature ^{1}H NMR technique, thus, permits estimation of $\Delta G^{\ddagger}$ for various polymers. As seen in the NMR spectra of poly(**9**) (Figure 8), two broad but clearly separated signals at 22 °C coalesce at 110 °C, and this process is reversible. From the coalesced temperature and the difference in the chemical shifts at 22 °C, $\Delta G^{\ddagger}$ is readily calculated to be 18.5 kcal/mol that is comparable to the value for polyisocyanates (*22*). Table III summarizes $\Delta G^{\ddagger}$ values for several poly(propiolic esters). From these results, the bulkiness of the pendant has proven to show no significant influence on the speed of the helix_helix transformation. Kinetic control of the helix is, therefore, probably impossible unless extremely bulky substituents are introduced to the pendant.

36

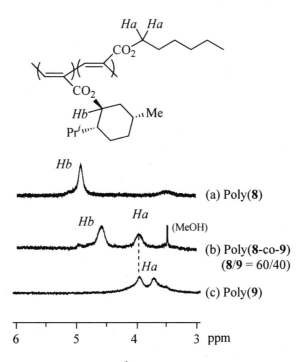

Figure 7. Variable temperature ^1H NMR (expanded) of poly(9) in toluene-d_6 (400 MHz).

(Reproduced with permission from reference 18. Copyright 2000 American Chemical Society.)

Table III. Activation Energy of Helix-Helix Interconversion for Poly(propiolic esters) $[(CH=CCO_2R)_n]$

Polymer (R =)	$M_n/10^{3\,a}$	$M_w/10^{3\,a}$	Solvent	$\Delta G^{\ddagger}$ (kcal/mol)
$n\text{-}C_6H_{13}$	78	325	toluene	18.5
$CH_2CH(CH_3)_2$	16	72	toluene	17.6
$CH_2C(CH_3)_3$	8.5	26	toluene	18.3
$CH_2CH_2CH(CH_3)_2$	140	248	$o\text{-}DCB^b$	19.0
CH_2CH_2Ph	100	197	toluene	17.5
$CH_2CH_2CH_2Ph$	30	136	$o\text{-}DCB^b$	19.3

[a] Estimated by GPC (PSt, THF). [b] o-Dichlorobenzene.

(Reproduced with permission from reference 18. Copyright 2000 American Chemical Society.)

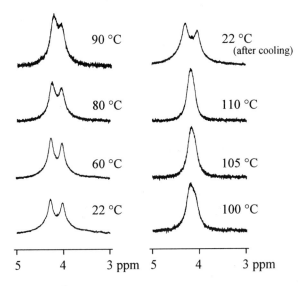

*Figure 8. Expanded ^1H NMR spectra of (a) poly(**8**), (b) copolymer of **8** with **9**
(**8**/**9** = 60 : 40) and (c) poly(**8**) in CDCl₃ (400 MHz).*
*(Reproduced with permission from reference 18. Copyright 2000 American
Chemical Society.)*

Free energy difference between the helical and helix reversal states (ΔG_r)

As stated above, the peak separation of the α-methylene signal is attributed
to the slow interconversion between two helices. The integrated intensities of
these two signals, thus, should be identical unless two methylene protons
possess unexpectedly different relaxation time. However, Figure 8 apparently
exhibits a difference in the intensity between the two signals. At every
temperature the peak at lower magnetic field has larger intensity, and an
increase in temperature leads to an increase in the difference in the peak
intensity. This is due to the overlapping of the signal attributed to the α-
methylene proton in disordered conformation. This idea is supported by the
fact that the resonance due to the α-methylene proton of randomly coiled
poly(**8**-*co*-**9**) (**8**/**9** = 60/40) is observed in the same region as the lower magnetic
field peak of the α-methylene signal in the helical state (Figure 7). The
temperature dependence of the variation of the integrated intensity ratio is
reasonably explained by the increasing population of the disordered state with
increasing temperature. The computational peak deconvolution of the
diastereotopic two signals into two Lorentzian functions allows the estimation of
the ratio of the helical and disordered states, which consequently gives the free
energy difference between these two states. The estimated value, 1.59±0.16
kcal/mol at 22 °C and 0.55±0.66 kcal/mol at 90 °C for the present sample of
poly(**9**), probably corresponds to ΔG_r because the disordered states can act as the

helix reversal state interposed between two helices with opposite screw-sense. The ΔG_r of poly(9) is smaller than that of polyisocyanates, which means that the persistence length of poly(propiolic esters) is smaller than that of polyisocyanates (*15, 23_27*).

Conclusion

This review article summarizes our recent efforts to comprehensively understand the nature of helical conformation of poly(propiolic esters). What should be emphasized is the fact that the two parameters, ΔG_r and $\Delta G^{\ddagger}$, conducting the stability of the helical structure, can be readily estimated by NMR technique. The profile of the helical conformation of poly(propiolic esters) is easily drawn as follows by using these parameters. Poly(propiolic esters) possess a higher population of helix reversal state than poly(isocyanates). However, the persistence length of the helix of poly(propiolic esters) is apparently larger than that of other polymers from monosubstituted acetylenes. Therefore, poly(propiolic esters) can exist in well-defined helical conformation in solution even if they do not possess bulky substituents. The speed of helix inversion is comparable to those of polyisocyanates. Inversion of the helix of poly(propiolic esters), thus, readily occurs at ambient temperature. Kinetic control of the helix sense probably requires extremely bulky pendant groups.

Acknowledgments

The authors greatly appreciate Dr. F. Kondo at Asahi Techneion Co., Ltd. for measurement of GPC equipped with viscometer. The authors are also indebted to Professor T. Yoshizaki at Kyoto University for many valuable suggestions. This work was supported by a Grant-in-Aid for Scientific Research on Priority Areas from the Ministry of Education, Science, Culture, and Sports, Japan.

References

1. Okamoto, Y.; Nakano, T. *Chem. Rev.* **1994**, *94*, 349_372.
2. Rowan, A. E.; Nolte, R. J. M. *Angew. Chem. Int. Ed. Engl.* **1998**, *37*, 63–68.
3. Qin, M.; Bartus, J.; Vogl, O. *Macromol. Symp.* **1995**, *12*, 387_402.
4. Nolte, R. J. M. *Chem. Soc., Rev.* **1994**, 11_19.
5. Okamoto, Y.; Hatada, K. In *Chromatographic Chiral Separations*; Zief, M., Crane, L. J., Eds.; Marcel Dekker: New York, 1988; p 199.
6. Pu, L. *Tetrahedron: Asymmetry* **1998**, *9*, 1457_147.

7. Pu. L. *Acta Polym.* **1997**, *48*, 116_141.

8. Ciardelli, F.; Lanzillo, S.; Pieroni, O. *Macromolecules* **1974**, *7*, 174_179.

9. Tang, B. Z.; Kotera, N. *Macromolecules* **1989**, *22*, 4388_4390.

10. Yamaguchi, M.; Omata, K.; Hirama, M. *Chem. Lett.* **1992**, 2261_2262.

11. Aoki, T.; Kokai, M.; Shinohara, K.; Oikawa, E. *Chem. Lett.* **1993**, 2009_ 2012.

12. Yashima, E.; Huang, S.; Okamoto, Y. *J. Chem. Soc., Chem. Commun.* **1994**, 1811_1812.

13. Yashima, E.; Huang, S.; Matsushima, T.; Okamoto, Y. *Macromolecules* **1995**, *28*, 4184_4193.

14. Kishimoto, Y.; Itou, M.; Miyatake, T.; Ikariya, T.; Noyori, R. *Macromolecules* **1995**, *28*, 6662_6666.

15. Green, M. M.; Gross, R. A.; Schilling, F. C.; Peterson, N. C.; Sato, T.; Teramoto, A.; Cook, R.; Lifson, S. *Science* **1995**, *268*, 1860_1866.

16. Nakako, H.; Nomura, R.; Tabata, M.; Masuda, T. *Macromolecules* **1999**, *32*, 2861_2864.

17. Nakako, H.; Mayahara, Y.; Nomura, R.; Tabata, M.; Masuda, T. *Macromolecules* **2000**, *33*, 3978–3982.

18. Nomura, R.: Fukushima, Y.; Nakako, H.; Masuda, T. *J. Am. Chem. Soc.* in press.

19. Tabata, M.; Inaba, Y.; Yokota, K.; Nozaki, Y. *J. Macromol. Sci., Pure Appl. Chem.* **1994**, *A31*, 465_475.

20. Masuda, T.; Kawai, M.; Higashimura, T. *Polymer* **1982**, *23*, 744_747.

21. Green, M. M.; Park, J-W.; Sato, T.; Teramoto, A.; Lifson, S.; Selinger, R. L. B.; Selinger, J. V. *Angew. Chem. Int. Ed.* **1999**, *38*, 3138–3154.

22. Ute, K.; Fukunishi, Y.; Jha, S. K.; Cheon, K. S.; Munoz, B.; Hatada, K.; Green, M. M. *Macromolecules* **1999**, *32*, 1304_1307.

23. Lifson, S.; Andreola, C.; Peterson, N. C.; Green, M. M. *J. Am. Chem. Soc.* **1989**, *111*, 8850_8858.

24. Gu, H.; Nakamura, Y.; Sato, T.; Teramoto, A.; Green, M. M.; Andreola, C.; Peterson, N. C.; Lifson, S. *Macromolecules*, **1995**, *28*, 1016_1024.

25. Okamoto, N.; Mukaida, F.; Gu, H.; Nakamura, Y.; Sato, T.; Teramoto, A.; Green, M. M. *Macromolecules* **1996**, *29*, 2878_2884.

26. Gu, H.; Nakamura, Y.; Sato, T.; Teramoto, A.; Green, M. M.; Jha, S. K.; Andreola, C.; Reidy, M. P. *Macromolecules* **1998**, *31*, 6362_6368.

27. Gu, H.; Sato, T.; Teramoto, A.; Varichon, L.; Green, M. M. *Polym. J.* **1997**, *29*, 77_84.

Chapter 4

Helicity Induction on Optically Inactive Polyacetylenes and Polyphosphazenes

Eiji Yashima[1,2] and Katsuhiro Maeda[1]

[1]Department of Molecular Design and Engineering, Graduate School of Engineering,
Nagoya University,
Nagoya, Japan
[2]Form and Function, PRESTO, JST, Chikusa-ku, Nagoya 464–8603, Japan

This paper discusses the helicity induction on optically inactive polymers bearing functional groups through non-covalent interactions with chiral small molecules. Polyphenylacetylenes and polyphosphazenes having carboxy groups are specific polymers for the induced helical polymers. They respond to the chirality of amines and exhibit an induced circular dichroism and a large optical rotation change, respectively, in the presence of optically active amines such as (R)-1-phenylethylamine. This indicates that the polymers may form a dynamic, one-handed helical conformation upon complexation with the chiral amine. Similar helicity induction on other optically inactive polymers is also briefly described.

Macromolecular helicity is the most important chirality which governs the sophisticated and fundamental properties of both biological and synthetic polymers. Therefore, significant attention has been paid to developing artificial helical polymers and oligomers with optical activity not only to mimic nature, but also to develop novel chiral materials in areas such as liquid crystals, membranes, and chiral selectors (1-10). Optically active helical polymers can be prepared by either the polymerization of optically active monomers or the asymmetric polymerization (screw-sense selective polymerization) of achiral or prochiral monomers with chiral catalysts or initiators (*1,7*). These helical polymers may be classified into two types with respect to the nature of the helical conformation: one is a stable helical polymer even in solution and the second is a dynamic helical polymer. Typical helical polymers exhibiting an optical activity due to a one-handed helicity are shown in Figure 1. Poly(triphenylmethyl methacrylate) (**1**), polychloral (**2**), polyisocyanides (**3**), and poly(2,3-quinoxalines) (**4**) belong to the stable helical polymers and the one-handed helical polymers can be prepared by the screw-sense selective polymerization.

Polyisocyanates (**5**) and polysilanes (**6**) are helical polymers with a long, alternate sequence of left- and right-handed helices, and equilibrium exists in solution between the helices separated by the helix reversal points that move along the polymer backbone (*2,3,10*), and therefore, they can be called dynamic helical polymers. The helix inversion barriers of the polymers are very low. Therefore, optically active polyisocyanates and polysilanes with a prevailing one-handed helix were successfully prepared through copolymerization of achiral monomers with a small amount of optically active monomers. This can be considered as a typical example of the chirality amplification in a polymer.

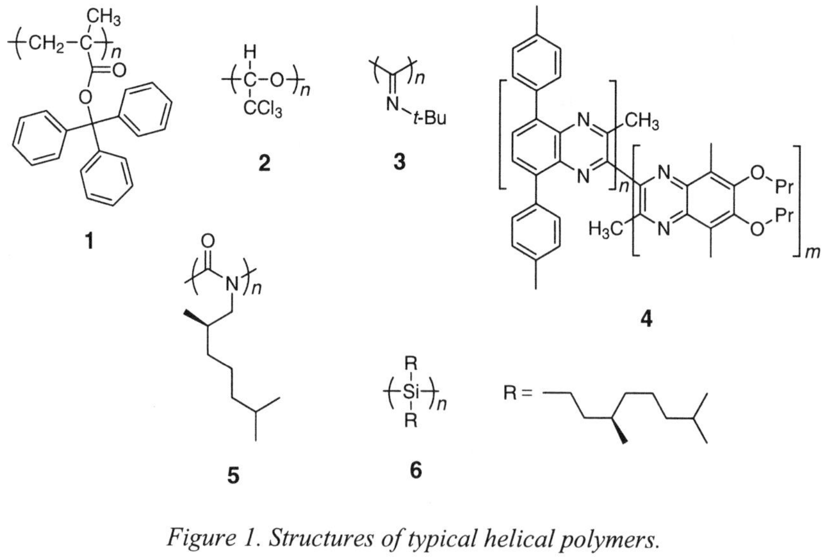

Figure 1. Structures of typical helical polymers.

However, we recently succeeded in inducing a conceptually new helicity on optically inactive polymers upon complexation with optically active small molecules capable of interacting with the functional groups of the polymers, which involve polyacetylenes and polyphosphazenes and show optical activity due to an induced macromolecular helicity. This chapter describes our recent studies on helicity induction of these polymers through non-covalent, chiral acid-base interactions.

Helicity Induction on Optically Inactive Polyacetylenes

Cis-transoidal, stereoregular polyphenylacetylenes bearing various functional groups, such as a carboxy (poly-**7**) (*11,12*), dihydroxyboro (poly-**8**) (*13*), or amino group (poly-**9**) (*14*) are specific examples of the induced helical polymers. These polymers are optically inactive and may be achiral, because they may have a large number of short helical units with many helix-reversal points. However, in the presence of optically active small molecules capable of interacting with these functional groups, for instance, chiral amines, sugars, and chiral acids, they form an induced one-handed helical structure (Figure 2) and show a characteristic induced circular dichroism (ICD) in the UV-visible region. The Cotton effect signs can be used as a novel probe for the assignments of their stereochemistry including the absolute configuration of the chiral molecules.

For example, poly-**7** forms a complex with various chiral amines and amino alcohols (**10**—**18**) through an acid-base interaction and the complexes showed a split-type ICD in the UV-visible region in both solution and the film state (Figure 3) (*11*). The split type of Cotton effects reflects the absolute configuration of the chiral amines; all primary amines and amino alcohols of the same configuration gave the same Cotton effect signs when the configurations are the same. The intensity of Cotton effects depends on the bulkiness of the chiral amines; the magnitude of the ICD likely increases with an increase in the bulkiness of the chiral amines, **14** < **13** < **12** « **11**, **10** as shown in Figure 3. The bulky groups introduced at the *para* position of poly-**7** may contribute more efficiently for the polymer to take a predominant screw-sense.

More interestingly, the induced helicity on poly-**7** was found to be able to be "*memorized*" when the chiral amine ((*R*)-**10**) is replaced by various achiral amines such as 2-aminoethanol, 3-amino-1-propanol, and *n*-butylamine (*12*). We have recently developed a simple and easy method for the preparation of *cis-transoidal* poly-**7** in water. In the presence of bases such as NaOH and amines, (4-carboxyphenyl)acetylene was rapidly and homogeneously polymerized with water-soluble rhodium complexes to yield yellow-orange fibrous, high molecular weight polymers in high yields. Moreover, the obtained sodium salt of poly-**7** responded to the chirality of natural amino acids and formed a predominantly one-handed helical structure in water, thus showing an ICD in the UV-visible region (*15*).

Similar helicity induction on an optically inactive polymer through acid-base interactions has been reported for polyguanidine (**19**) (*16*), polyisocyanate (**20**) (*17*), polyaniline (**21**) (*18*), and polypyrrole (**22**) (*19*). For instance, Novak, et al. reported an irreversible transition for an optically inactive polyguanidine

44

(19) during the complexation with (S)-camphorsulfonic acid (23) at higher temperatures. They observed dramatic increase in optical rotation after annealing the mixture (16).

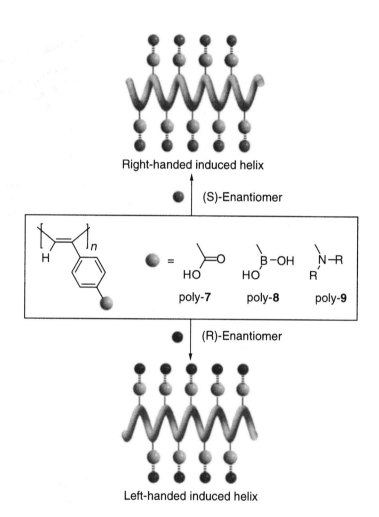

Figure 2. Schematical illustrations of the one-handed helicity induction on the optically inactive polyphenylacetylenes bearing a functional group with optically active compounds.

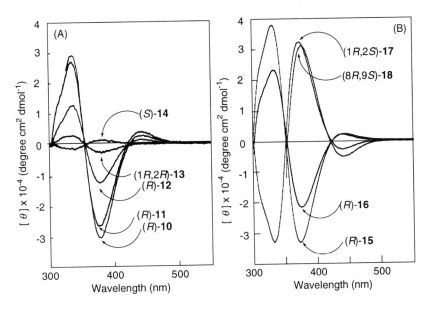

*Figure 3. CD spectra of poly-7 with (R)- or (S)-amines (**10—14**) (A) and amino alcohols (**15—18**) (B) in DMSO. The CD spectra were measured in a 0.05 cm quartz cell at room temperature (ca. 20—22 °C) with a poly-7 concentration of 1.0 mg (6.8 μmol monomer units)/mL. [amine]/[poly-7] = 50. (Reproduced from ref. 11. Copyright 1997 American Chemical Society).*

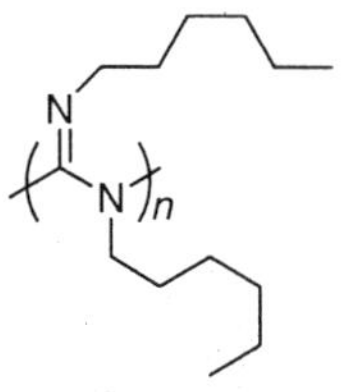

19

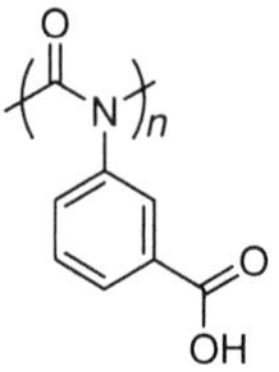

20

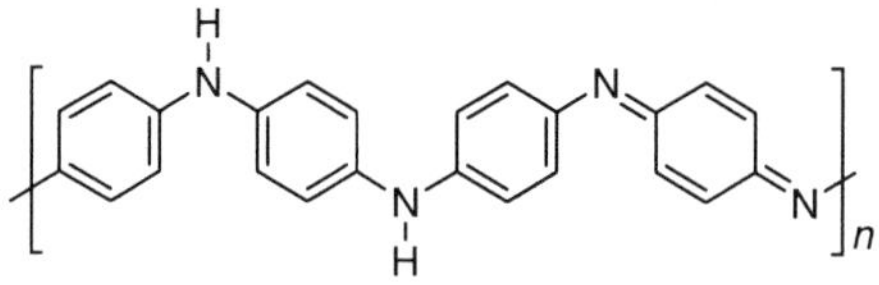

21: Emeraldine base (EB)

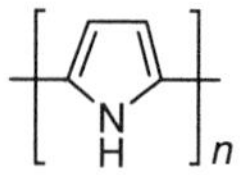

22

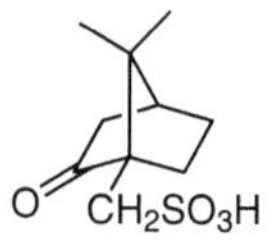

23

As described in the previous section, a one-handed helical conformation can be induced on optically inactive polymers such as stereoregular, functional polyacetylenes in the presence of chiral small molecules. We applied this methodology to an optically inactive poly(organophosphazene) bearing carboxy groups and examined whether the polymer could form a predominantly one-handed helical conformation in solution by responding to the chirality of optically active amines (*20*).

Synthesis of Poly(organophosphazenes)

Polyphosphazenes are useful polymers as technologically valuable, new polymeric materials because they have attractive properties such as high thermal stability, flame retardant, low temperature flexibility, and biocompatibility in some cases (*21-27*). The classical, but still useful method to prepare poly(organophosphazenes) developed by Allcock involves the thermal ring opening polymerization of hexachlorocyclotriphosphazene (**24**) followed by the macromolecular substitution reaction with nucleophiles such as alkoxides and primary amines (eq 1). Various kinds of poly(organophosphazenes) can be readily prepared from a single polyphosphazene, the poly(dichlorophosphazene) (**25**), and their properties can be widely tuned by proper selection of the substituents introduced in the polymers.

Recently, Allcock and co-workers discovered the living cationic-induced polymerization of trichloro(trimethylsilyl)phosphoranimine (**26**) with trace amounts of PCl$_5$ to yield **25** with well-defined molecular weights and narrow molecular weight distributions (eq 2) (*28,29*). We used this living

$$\text{(1)}$$

$$\text{LiN(SiCH}_3)_2 + \text{PCl}_5 \longrightarrow \text{Cl}-\overset{\overset{\displaystyle Cl}{|}}{\underset{\underset{\displaystyle Cl}{|}}{P}}=\text{N}-\text{Si(CH}_3)_3 \xrightarrow[\text{CH}_2\text{Cl}_2,\ 30\ °\text{C}]{\text{PCl}_5} \left(\text{N}=\overset{\overset{\displaystyle Cl}{|}}{\underset{\underset{\displaystyle Cl}{|}}{P}}\right)_n \quad \text{(2)}$$

26 **25**

polymerization method to prepare a poly(organophosphazene) bearing carboxy groups.

Structures of Polyphosphazenes

Conformation of the polyphosphazenes including **25** and other substituted poly(organophosphazenes) such as poly(bis(phenoxy)phosphazene) has been extensively studied using X-ray diffraction analyses (*30-37*) and computational calculations based on empirical, semiempirical, and the *ab initio* method (*36,38-43*). However, there are controversies about these results and both the helical (*30,31,38,39*) and *cis-trans* planar (*32-36,40-43*) conformations have been proposed for the polyphosphazenes. On the other hand, a number of optically active polyphosphazenes bearing chiral side groups, such as amino acid esters (**27, 28**) (*44,45*) and glucosyl (**29**) (*46,47*) and steroidal groups (**30**) (*48*), have been prepared in order to develop biocompatible and biodegradable materials, but their conformations have not yet been examined in detail.

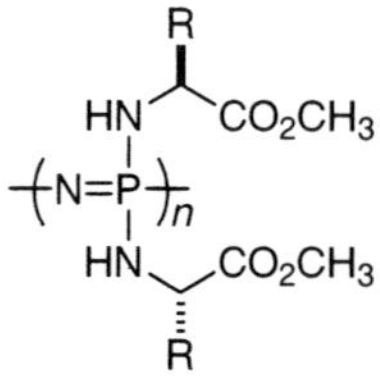

R = CH$_3$, CH$_2$CH(CH$_3$)$_2$, CH$_2$Ph

27

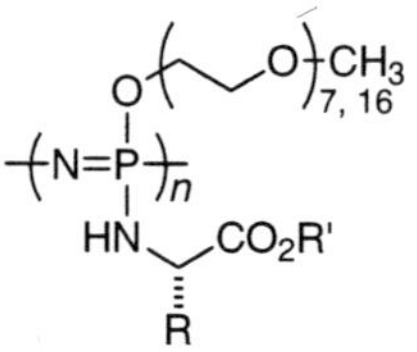

R = CH$_3$, CH$_2$CH(CH$_3$)$_2$, CH$_2$Ph
R' = CH$_3$, C$_2$H$_5$, CH$_2$Ph

28

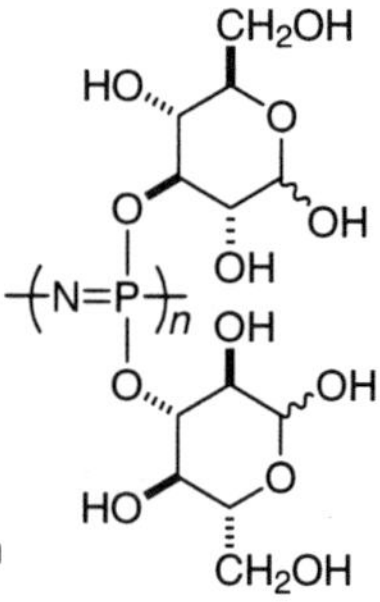

29

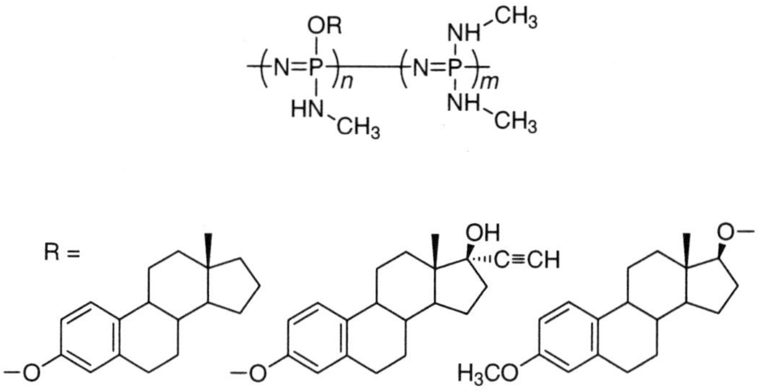

R =

30

Helicity Induction on Poly(bis(4-carboxyphenoxy)phosphazene)

Poly(bis(4-carboxyphenoxy)phosphazene) (poly-**31**) was prepared according to the reported method as shown in eq 3 (*49-51*). The molecular weight (*Mn*) and the distribution (*Mw/Mn*) of poly-**31** were estimated to be 1.4 x 10^5 (degree of polymerization, *ca.* 370) and 1.2, respectively, by size exclusion chromatography (SEC) as its ethyl esters. A model compound of poly-**31**, hexakis(4-carboxyphenoxy)cyclotriphosphazene (**32**) was also prepared for comparison (*20*).

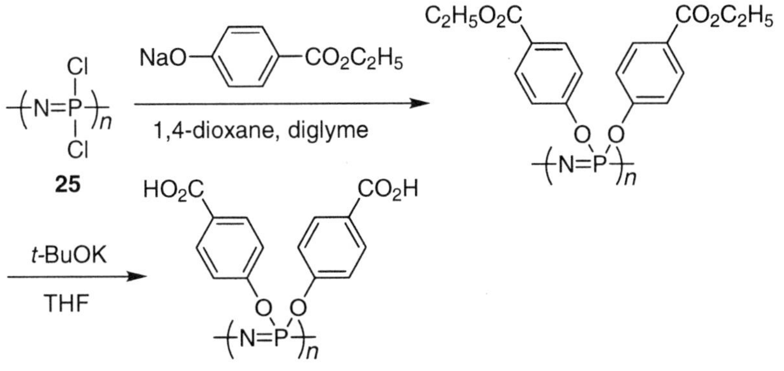

poly-**31**

(3)

The chiroptical properties of poly-**31** were investigated in DMSO in the presence of an optically active amine, (R)-1-phenylethylamine ((R)-**12**) and a polarimeter was used to detect the changes in the optical rotation of the complex ([poly-**31**] = 0.5 g/dL, [(R)-**12**]/[carboxy residues of poly-**31**] = 5) in the temperature range of 20 — 65 °C (Figure 4). The optical rotation values were calculated based on the poly-**31** concentration. Polyphosphazenes show no UV or visible absorption spectra at wavelengths longer than 220 nm due to the main chains (*21*), and therefore, CD, a very powerful technique to detect chiral polymer conformations, could not be used.

The optical rotation of the poly-**31**—(R)-**12** complex in DMSO slowly changed from a positive value to a negative one and increased in the negative direction with an increase in temperature. After annealing the complex at 65 °C for *ca.* 2 h, the optical rotation exhibited a large negative value ($[\alpha]_D^{65}$ −127°). The sign of the rotation was opposite to that of (R)-**12** ($[\alpha]_D^{25}$ +33° in DMSO). This first-annealing process was irreversible, but after the solution was once annealed at 65 °C, the poly-**31** became reversible with further rapid optical rotational changes in the temperature range of 20 — 65 °C and showed a larger optical rotation ($[\alpha]_D^{20}$ −197° in DMSO).

On the other hand, the optical rotation of the cyclic trimer (**32**)—(R)-**12** complex hardly changed in the same temperature ranges and the optical rotation sign was the same as that of (R)-**12**, but was opposite to that of the annealed poly-**31**—(R)-**12** complex (Figure 4). The net specific rotation of the **32**—(R)-**12** complex is almost exactly the same as that of pure (R)-**12**, indicating that any chirality was induced on the cyclic trimer **32** in the presence of (R)-**12**.

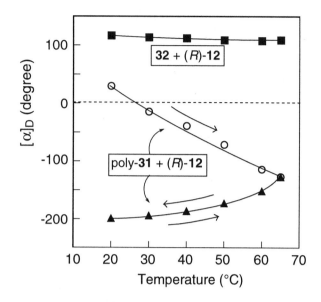

Figure 4. Optical rotation changes in the poly-31— (O, ▲) and 32—(R)-12 (■) complexes ([(R)-12]/[carboxy residues of poly-31 or 32] = 5) in DMSO at various temperatures. The rotations were first measured at 20, 30, 40, 50, 60, and 65 °C after annealing the samples at these temperatures for ca. 2 h (first-annealing (O)). During the first-cooling and further annealing and cooling processes (▲), the rotations were measured after allowing the sample to stand for ca. 10 min at these temperatures. (Reproduced from ref. 20. Copyright 2000 American Chemical Society).

These results suggest that the large negative optical rotation of the poly-**31**—(*R*)-**12** complex may be derived from the helicity induction with a one handedness in excess on the polymer main chain upon complexation with the optically active (*R*)-**12** through the acid base interaction, as observed in the complex formations of poly((4-carboxyphenyl)acetylene) (poly-**7**) with chiral amines as described in the previous section. When the enantiomeric amine, (*S*)-**12** was used instead of (*R*)-**12**, the poly-**31**—(*S*)-**12** complex showed mirror image optical rotational changes. Recently, Carriedo et al. reported that the poly(organophosphazene) bearing an optically active binaphthyl group (**33**) might adopt a helicoidal secondary structure in solution based on their optical rotation measurements (*52*). The polymer **33** showed larger optical rotation than that of the corresponding model cyclic trimer.

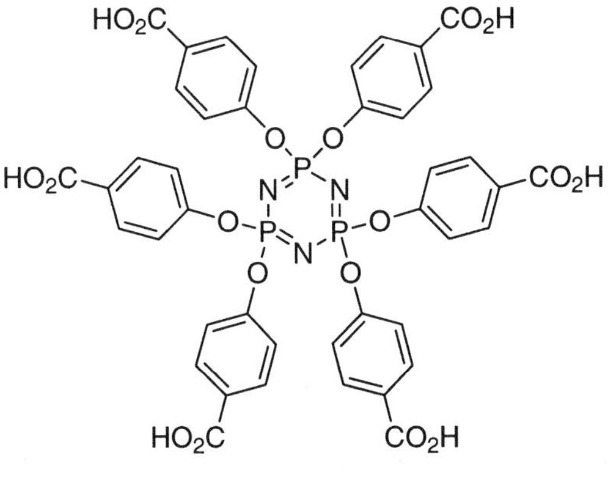

32

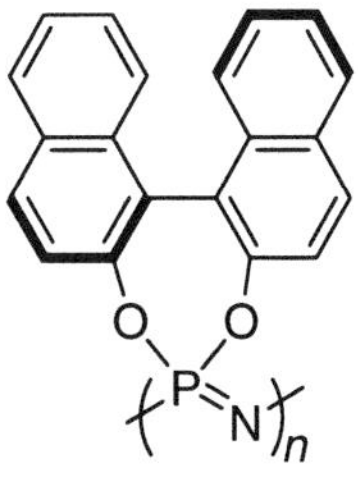

33

When an equimolar amount of strong acid, such as trifluoroacetic acid, was added to the poly-**31**—(*R*)-**12** complex solution, the optical rotation instantly changed to the same value as that of the cyclic trimer **32**—(*R*)-**12** complex, because trifluoroacetic acid preferentially complexes with the optically active amine so that the poly-**31** reverts to the original, optically inactive polymer. This suggests that the original, free poly-**31** can not maintain the induced helical conformation in solution, but a dynamic, one-handed helical conformation can be induced upon complexation with the optically active amines.

Moreover, we found that the optical rotation of the complex decreased when the sample was annealed at higher temperatures (> 80 °C), but after further annealing the samples at 65 °C for 2 h again, the specific rotations recovered to their original values ($[\alpha]_D^{20}$ *ca.* –200°). We believe that these dynamic changes in the optical rotation of the poly-**31**—(*R*)-**12** complex also support that the poly-**31** may have a one-handed helical conformation in the presence of the chiral amines.

The structure for the induced helical conformation of the poly-**31**—(*R*)-**12** complex is not clear at the present time, but the twisted *cis-trans* (*38*) or helical *cis-gauche-gauche* conformation (*39*) may be specific to the induced helical conformation.

The results described in this chapter may indicate that the introduction of chiral substituents to the polymer backbone through covalent bonding is not necessary for induction of the main chain helical chirality onto the polymer backbone. Non-covalent interactions may be able to induce a dynamic, helical conformation in polymers (*53*).

52

Acknowledgment.

This work was supported in part by Grant-in-Aid for Scientific Research from the Ministry of Education, Science, Sports, and Culture, Japan and the Daiko Foundation (E.Y.).

Literature Cited

(1) Okamoto, Y.; Nakano, T. *Chem. Rev.* **1994**, *94*, 349-372.
(2) Fujiki, M. *J. Am. Chem. Soc.* **2000**, *122*, 3336-3343.
(3) Green, M. M.; Peterson, N. C.; Sato, T.; Teramoto, A.; Cook, R.; Lifson, S. *Science* **1995**, *268*, 1860-1866.
(4) Bassani, D. M.; Lehn, J.-M. *Bull. Soc. Chim. Fr.* **1997**, *134*, 897-906.
(5) Seebach, D.; Matthews, J. L. *Chem. Commun.* **1997**, 2015-2022.
(6) Pu, L. *Acta Polym.* **1997**, *48*, 116-141.
(7) Rowan, A. E.; Nolte, R. J. M. *Angew. Chem. Int. Ed.* **1998**, *37*, 63-68.
(8) Gellman, S. H. *Acc. Chem. Res.* **1998**, *31*, 173-180.
(9) Prince, R. B.; Barnes, S. A.; Moore, J. S. *J. Am. Chem. Soc.* **2000**, *122*, 2758-2762.
(10) Green, M. M.; Park, J.-W.; Sato, T.; Teramoto, A.; Lifson, S.; Selinger, R. L. B.; Selinger, J. V. *Angew. Chem. Int. Ed.* **2000**, *38*, 3138-3154.
(11) Yashima, E.; Matsushima, T.; Okamoto, Y. *J. Am. Chem. Soc.* **1997**, *119*, 6345-6359.
(12) Yashima, E.; Maeda, K.; Okamoto, Y. *Nature* **1999**, *399*, 449-451.
(13) Yashima, E.; Nimura, T.; Matsushima, T.; Okamoto, Y. *J. Am. Chem. Soc.* **1996**, *118*, 9800-9801.
(14) Yashima, E.; Maeda, Y.; Matsushima, T.; Okamoto, Y. *Chirality* **1997**, *9*, 593-600.
(15) Saito, M. A.; Maeda, K.; Onouchi, H.; Yashima, E. *Macromolecules* **2000**, *33*, 4616-4618.
(16) Schlitzer, D. S.; Novak, B. M. *J. Am. Chem. Soc.* **1998**, *120*, 2196-2197.
(17) Maeda, K.; Yamamoto, N.; Okamoto, Y. *Macromolecules* **1998**, *31*, 5924-5926.
(18) Majidi, M. R.; Kane-Maguire, L. A. P.; Wallace, G. G. *Polymer* **1995**, *36*, 3597-3599.
(19) Zhou, Y.; Zhu, G. *Polymer* **1997**, *38*, 5493-5495.
(20) Yashima, E.; Maeda, K.; Yamanaka, T. *J. Am. Chem. Soc.* **2000**, *122*, 7813-7814.
(21) Allcock, H. R. *Chem. Rev.* **1972**, *72*, 315-356.
(22) Allcock, H. R. *Science* **1976**, *193*, 1214-1219.
(23) Allcock, H. R. *Angew. Chem. Int. Ed. Engl.* **1977**, *16*, 147-156.
(24) Mark, J. E; Allcock, H. R.; West, R. *Inorganic Polymers*; Prentice Hall: Englewood Cliffs, NJ, 1992.

(25) Manners, I. *Angew. Chem. Int. Ed. Engl.* **1996**, *35*, 1602-1621.

(26) White, M. L.; Matyjaszewski, K. *Macromol. Chem. Phys.* **1997**, *198*, 665-671.

(27) Jaeger, de R.; Gleria, M. *Prog. Polym. Sci.* **1998**, *23*, 179-276.

(28) Honeyman, C. H.; Manners, I.; Morrissey, C. T.; Allcock, H. R. *J. Am. Chem. Soc.* **1995**, *117*, 7035-7036.

(29) Allcock, H. R.; de Denus, C. R.; Prange, R.; Nelson, J. M. *Macromolecules* **1999**, *32*, 7999-8004.

(30) Meyer, K. H.; Lotmar, W.; Pankow, G. W. *Helv. Chim. Acta* **1936**, *19*, 930-948.

(31) Giglio, E.; Pompa, F.; Ripamonti, A. *J. Polym. Sci.* **1962**, *59*, 293-300.

(32) Allcock, H. R.; Kugel, R. L.; Valan, K. J. *Inorg. Chem.* **1966**, *5*, 1709-1715.

(33) Bishop, S. M.; Hall, I. H. *Br. Polym. J.* **1974**, *6*, 193-204.

(34) Allcock, H. R.; Arcus, R. A.; Stroh, E. G. *Macromolecules* **1980**, *13*, 919-928.

(35) Allcock, H. R.; Tollefson, N. M.; Arcus, R. A.; Whittle, R. R. *J. Am. Chem. Soc.* **1985**, *107*, 5166-5177.

(36) Caminiti, R.; Gleria, M.; Lipkowitz, K. B.; Lombardo, G. M.; Pappalardo, G. C. *J. Am. Chem. Soc.* **1997**, *119*, 2196-2204.

(37) Miyata, T.; Yonetake, K.; Masuko, T. *J. Mater. Sci.* **1994**, *29*, 2467-2473.

(38) Boehm, R. C, *J. Phys. Chem.* **1993**, *97*, 13877-13886.

(39) Sun, H. *J. Am. Chem. Soc.* **1997**, *119*, 3611-3618.

(40) Allcock, H. R.; Allen, R. W.; Meister, J. J. *Macromolecules* **1976**, *9*, 950-955.

(41) Allen, R. W.; Allcock, H. R. *Macromolecules* **1976**, *9*, 956-961.

(42) Tanaka, K.; Yamashita, S.; Yamabe, T. *Macromolecules* **1986**, *19*, 2062-2064.

(43) Bougeard, D.; Br mard, C.; Jaeger, R. D.; Lemmouchi, Y. *J. Phys. Chem.* **1992**, *96*, 8850-8855.

(44) Allcock, H. R.; Fuller, T. J.; Mack, D. P.; Matsumura, K.; Smeltz, K. M. *Macromolecules* **1977**, *10*, 824-830.

(45) Song, S.-C.; Lee, S. B.; Jin, J.-I.; Sohn, Y. S. *Macromolecules* **1999**, *32*, 2188-2193.

(46) Allcock, H. R.; Scopelianos, A. G. *Macromolecules* **1983**, *16*, 715-719.

(47) Allcock, H. R.; Pucher, S. R. *Macromolecules* **1991**, *24*, 23-34.

(48) Allcock, H. R.; Fuller, T. J. *Macromolecules* **1980**, *13*, 1338-1345.

(49) Allcock, H. R.; Kwon, S. *Macromolecules* **1989**, *22*, 75-79.

(50) Allcock, H. R.; Crane, C. A.; Morrissey, C. T.; Nelson, J. M.; Reeves, S. D.; Honeyman, C. H.; Manners, I. *Macromolecules* **1996**, *29*, 7740-7747.

(51) Stewart, F. F.; Lash, R. P.; Singler, R. E. *Macromolecules* **1997**, *30*, 3229-3233.

(52) Carriedo, G. A.; Garc a-Alonso, F. J.; Gonz lez, P. A.; Garc a-Alvarez, J. L. *Macromolecules* **1998**, *31*, 3189-3196.

(53) Yashima, E.; Okamoto, Y. In *Circular Dichroism 2nd Ed.*; Berova, N.; Nakanishi, K.; Woody, R. W., Eds.; Wiley-VCH: New York, NY, 2000; Chapter 18.

Chapter 5

Helical and Higher Structural Ordering in Poly(3-methyl-4-vinylpyridine)

Lisandra J. Ortiz[1], Lawrence Pratt[2], Kimberly Smitherman[2], Biswajit Sannigrahi[2], and Ishrat M. Khan[1,*]

[1]Department of Chemistry, Clark Atlanta University, 223 James P. Brawley Drive, S.W., Atlanta, GA 30314
[2]Department of Chemistry, Fisk University, Nashville, TN 37208

Optically active (+) and (-) poly(3-methyl-4-vinylpyridine) have been prepared using chiral anionic initiating complexes at -78^0 C in toluene. At -4° C, the optical activity of both the (+) and the (-) poly(3-methyl-4-vinylpyridine) decreased with time to zero rotation. The polymers are optically stable (i.e. no change in optical activity) in the solid state at room temperature. These observations favor that higher structural order, most likely helical, is responsible for the observed optical activity of the polymers. Molecular modeling (MM3 and PM3) and conformational studies suggest that the secondary structure is a helix. The helicity selection of a single-screw sense via "chaperoning" starting with a racemic mixture of helical poly(3-methyl-4-vinylpyridine) may be carried out using chiral acid-base interactions and stable ICD (induced circular dichroism) spectra were observed.

Introduction

The preparation of synthetic macromolecules with higher structural order is of interest because of potential application of such macromolecules in material science, sensor technology, chiral synthesis and separation, and as biomaterials *(1,2)*. The conformationally stable helix may be considered as the simplest higher structurally ordered macromolecule. Several groups have reported the preparation of one-handed helical synthetic polymeric structures from achiral monomers such as bulky methacrylates, chloral, isocyanides, and isocyanates *(2,3)*. The most comprehensively studied and understood system of conformationally stable helical polymers are the bulky polymethacrylates developed by Okamoto and coworkers *(2)*. Among, these bulky polymers, poly(triphenylmethyl methacrylate) has been demonstrated to exhibit excellent chiral recognition and has been utilized as chiral stationary phase in high-performance liquid chromatography *(4)*. The reactive or labile ester groups in the methacrylate, however, is susceptible to hydrolysis during application as chiral stationary phase or as supports in asymmetric synthesis *(5)*. Thus, it is of interest to develop novel conformationally stable helical structures, which are relatively easy to prepare and do not carry reactive functional groups or, if they carry functional groups, the groups must be stable during the particular application. Additionally, the development of facile synthetic approaches for preparing stable helical polymers will provide easy access to secondary structures, which may be utilized as building blocks for preparing supra-molecular assemblies with tertiary structures *(6,7)*.

Recent developments in preparation of higher structurally ordered, most likely helical, synthetic polymers using relatively simple vinyl monomeric structures, such as 3-methyl-4-vinylpyridine *(8)* and 2-[(S)-2-(1-pyrrodinylmethyl)-1-pyrrolidinylmethyl]styrene *(9)*, suggests that preparation of other structurally similar conformationally ordered polymers is a real possibility. In order to understand the types of less bulky vinyl monomers which may be successfully polymerized to give optically active polymers, where the optical activity is due to a higher structural order, let us examine the rationale for selecting 3-methyl-4-vinylpyridine as a monomer for preparing conformationally ordered polymers. It is well known that in the solid state isotactic α-olefins and certain substituted polystyrenes are present in a low energy helical conformation with trans-gauche (...TGTG...) type backbone conformation *(10)*. The question is, whether it is possible to prepare synthetic helical (or higher ordered structured) polymers using similar types of monomers? In addition, the activation energy of helix-to-helix interconversion or helix-coil transition should be sufficiently large so that the conformationally stable structures, once prepared, are stable in solution. The selection of the monomer, 3-methyl-4-vinylpyridine, was further motivated by the reported preparation of isotactic poly(3-methyl-2-

vinylpyridine) by anionic polymerization *(11)*. An important observation in the poly(3-methyl-2-vinylpyridine) report was that isotacticity of this polymer plausibly resulted from the formation of a favored (low-energy) helical conformation caused by the non-bonded interaction between the 3-methyl group and the penultimate pyridine group i.e. it seems to indicate that the ortho methyl group is crucial to favoring the formation of a helical conformation. Motivated by these observations, the anionic polymerization of 3-methyl-4-vinylpyridine was carried out to successfully obtain polymers with controlled secondary conformations, most likely helical *(8)*. The recent report by the Okamoto group on the preparation of higher structurally ordered poly[2-[(S)-2-(1-pyrrodinylmethyl)-1-pyrrolidinylmethyl]styrene, supports this hypothesis *(9)*.

Results and Discussion

The preparation of (+) and (-), most likely, helical poly(3-methyl-4-vinylpyridine), abbreviated as P3M4VP, by anionic polymerization has been reported *(8)*. The helix sense selective polymerization was carried out by the method shown in Scheme 1. Using the (+) DDB and the (-) DDB as the chiral ligand optically polymers with (-) and (+) optical rotations, respectively, were prepared. At -4 °C, the optical rotation of both the (+) and the (-) polymers decreased with time to zero rotation. The loss of optical activity is most likely because of helix-to-helix interconversion resulting in racemization. While in solution at -4°C, mutarotation is observed, the polymers are stable (i.e. no change in optical activity) in the solid state at room temperature. The polymers may be stored in a refrigerator for several months without any loss of optical activity. Helix-to-helix interconversion is not observed in solution at -78°C. These observations favor that some sort of higher structure, most likely secondary helical structure, is giving rise to the optical activity. Because of the rapid helix-to-helix interconversion in solution at room temperature and the time required to obtain a CD spectra, it has not been possible to obtain CD spectra of the (+) and (-) polymers in the solution. The conformation of (+) and (-) P3M4VP, however, may be locked in an elastomeric solid matrix of poly(ethylene oxide) (PEO, MW 600,000) and poly(ethylene glycol) (PEG, MW 3000). The (+) and (-) P3M4VP/PEO (MW 600,000)/PEG (MW 3000) [1:1:1 by weight ratio] composites were prepared in the following manner. Helix sense selective polymerizations were carried out to produce 0.5 grams of the (+) or the (-) P3M4VP at -78° C. The living polymerizations were terminated and 0.5 grams of PEO and 0.5 grams of PEG were immediately added to the polymer solution at -78° C. The polymeric mixture was coprecipitated into cold hexanes, chilled with ice, to form the composites. The composites were dried in a vacuum oven at room temperature. The CD spectra of the composites show mirror image Cotton

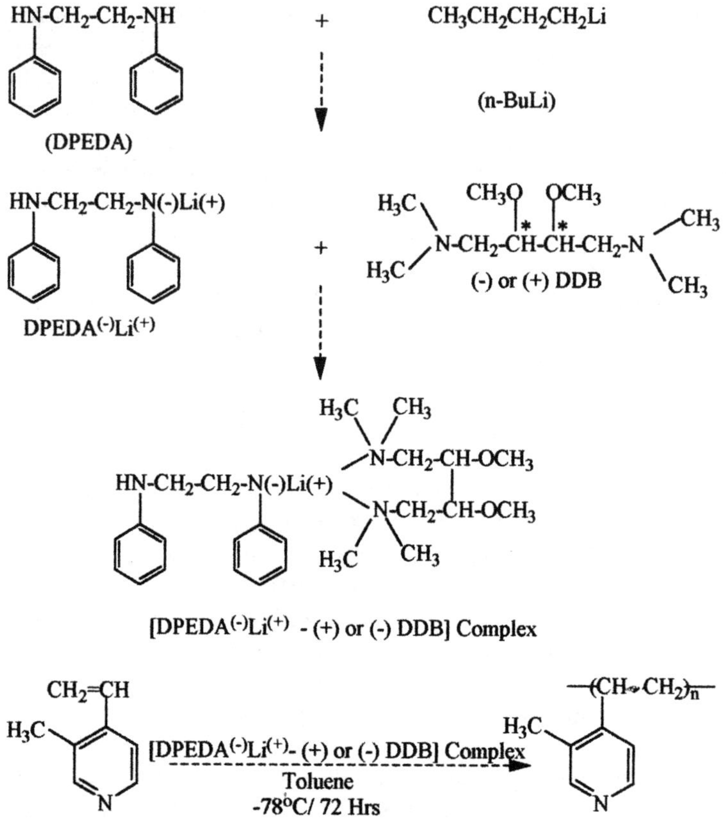

HN-CH$_2$-CH$_2$-NH + CH$_3$CH$_2$CH$_2$CH$_2$Li

(n-BuLi)

(DPEDA)

HN-CH$_2$-CH$_2$-N(-)Li(+) +

H$_3$C, CH$_3$O OCH$_3$
N-CH$_2$-CH-CH-CH$_2$-N—CH$_3$
H$_3$C (-) or (+) DDB CH$_3$

DPEDA$^{(-)}$Li$^{(+)}$

H$_3$C CH$_3$
N-CH$_2$-CH-OCH$_3$
HN-CH$_2$-CH$_2$-N(-)Li(+)
N-CH$_2$-CH-OCH$_3$
H$_3$C CH$_3$

[DPEDA$^{(-)}$Li$^{(+)}$ - (+) or (-) DDB] Complex

CH$_2$=CH
H$_3$C [DPEDA$^{(-)}$Li$^{(+)}$- (+) or (-) DDB] Complex —(CH—CH$_2$)$_n$—
H$_3$C
Toluene
-78°C/ 72 Hrs

Scheme 1. Synthesis of Optically Active Helical Polymers (Reproduced from ref. 8. Copyright 1998 American Chemical Society)

effect signals at 212 nm and 223 nm, indicating the formation of enantiomeric higher structural order. The helical conformation of the (+) and the (-) P3M4VP is stable in the PEO/PEG matrix at room temperature. Therefore, even in instances where the activation energy of the helix-to-helix interconversion is small such that the left- and the right-handed helices are rapidly equilibrating in solution, it is possible to readily lock in the helical conformation in a solid matrix. In the solid composite, the helix-to-helix interconversion process is not a molecular process but a process that requires the segmental motion or rearrangement of the neighboring polymer chains such that conformational reorientation required for helix-to-helix interconversion is possible. When the (+) and (-) P3M4VP/PEO/PEG composites are heated to 60 °C for 30 minutes, the Cotton effect signals are lost indicating either helix-to-helix interconversion resulting in racemization or helix-to-coil transition resulting in loss of the helical structure *(12)*. These observations suggest that the optically activity arises because of secondary structure and is not due to the presence of chiral center(s) in the polymer chain.

What is the secondary structure that is responsible for the observed optical activity? The observation by Natta and coworkers that ortho substituted isotactic polystyrenes in the solid state are in fact present in helical conformations, supports the idea that the secondary structure is helical and this idea is reinforced by the reported helical conformation of poly(3-methyl-2-vinylpyridine) *(11)*. Molecular modeling studies have been carried out to determine the preferred conformation of poly(3-methyl-4-vinypyridine). Oligomers of 3-methyl-4-vinylpyridine were modeled using the Spartan series of programs. Molecular mechanics calculations were performed with MM3, and optimized geometries were further optimized with the PM3 semi-empirical method. Models were built of the mm triads, mmm tetrads and mmmm pentads, in both helical and non-helical form, and the geometries were optimized by MM3 and PM3. Both MM3 and PM3 calculations support that helical isotactic tetramers and pentamers are more stable than the non-helical forms. The results of the calculations are shown in table 1.

Table 1: Calculated energy (Kcal/mole) of 3-methyl-4-vinylpyridine oligomers

Oligomer	Strain Energy (MM3)	H_f (PM3)
mmm helix	75.7	54.2
mmm non-helix	83.6	60.8
mmmm helix	98.2	75.5
mmmm non-helix	100.1	80.5

Only small differences in energy were found for the different isomers of the 3-methyl-4-vinylpyridine dimers and trimers, as the end groups were able to undergo conformation changes that affected the optimized energy of the entire molecule. Additionally, no stable helix was found for the rrr tetrad, which supports our premise that poly(3-methyl-4-vinylpyridine) has an isotactic configuration.

Systematic conformation search was carried out on two isotactic enantiomeric oligomers, (2R,4S,6R,8S,10R,12S)-2,4,6,8,10,12-hexa(3-methyl-4-pyridyl)tetradecane, abbreviated as *Enantiomer A* and (2S,4R,6S,8R,10S,12R)-2,4,6,8,10,12-hexa(3-methyl-4-pyridyl)tetradecane, abbreviated as *Enantiomer B*, using the Alchemy Series of programs *(13)*. The conformation search results

60

show the presence of a low energy left-handed helical conformation for *enantiomer B* and a low energy right-handed helical conformation for *enantiomer A*, (See Figure 1) i.e. shows a correlation between the enantiomeric secondary structure with the primary structure. The conformation search indicates that the helical twist starts with the third repeat unit in the oligomer. The systematic conformation search was carried out by specifying the number of rotatable bonds (n bonds) and the rotation increment in degrees (delta) in the oligomers. The systematic conformation search examined the total number of conformations, N and since only one increment to all rotatable bonds was set, N is calculated by the following expression:

$$N = [360/delta]^{n\ bonds}$$

For each conformation achieved by this set of rotations, all internal atomic distances along with the potential energy of a conformation are computed. The conformational studies suggest that it is reasonable to conclude that the secondary structure of isotactic poly(3-methy-4-vinylpyridine) is most likely helical.

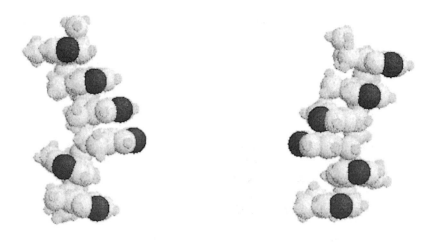

Figure 1: Left-handed and right-helical conformations of enantiomers B and A, respectively. [From left to right]

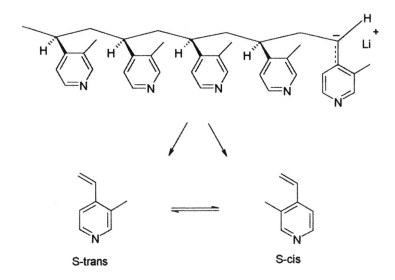

S-trans   S-cis

Scheme 2. Possible modes of monomer addition to living poly(3-methyl-4-vinylpyridine)

Conformational analysis suggests that the orientation of the 3-methyl group in the polymer chain is very important for the helical conformation to be stable. It is necessary for the methyl groups to be oriented in the same direction. This is shown in scheme 2. If the methyl groups were randomly oriented, the helical conformations were less favorable. This observation also points to the possible mechanism for the mode of the monomer addition to the living polymer chain end. Addition of the monomer in a conformation close to the S-trans would maintain the methyl group orientation, i.e. same direction, and hence the growing helical conformation. MM3 calculations indicate that monomer conformations close to S-trans are significantly lower in energy than conformations close to S-cis. It would be interesting to study the intermediate carbanion of living poly(3-methyl-4-vinylpyridine). Monomer addition in the S-trans mode should result in the formation of an E-isomer for the intermediate carbanion.

The relatively low activation energy of the helix-to-helix interconversion may permit helicity enrichment via chiral "chaperoning". Complexation of a racemic helical mixture with enantiomeric small molecules should result in the formation of a diastereomeric pair and a "chaperoned" shift of the equilibrium towards the lower energy diastereomer. "Chaperoning" by acid-base interaction has been reported to be effective for helicity selection and induction. Achiral poly(isocyanates), poly(acetylenes), poly(guanidines), and poly(organophosphazene)s have been "chaperoned" or induced to a helical conformation using chiral acid-based interactions *(14,15)*. To study this possible helicity selection by "chaperoning", a number of complexes of the poly(3-methyl-

4-vinylpyridine) with (R)- and (S)-mandelic acids were prepared and the ICD (induced circular dichroism) spectra of the complexes were recorded in the solid state. Complexes of racemic helical P3M4VP/(R)- and (S)-mandelic acid, abbreviated as MA, were prepared via acid-base interaction in tetrahydrofuran (THF), H_2O, and CH_3OH. The P3M4VP/(R)- or (S)-mandelic acid complexes were prepared by stirring 0.5 grams of P3M4VP with the appropriate amount of the (R)- or (S)-mandelic acid as for 8 hours at room temperature. The solution was poured onto poly(ethylene) plates and dried under vacuum to obtain the CD samples. The weight average molecular weights (M_w) of the polymers, relative to polystyrene standards, used in preparing the complexes were between 30,000 and 35,000. The $[\alpha]^{20}_D$ of (R)- and (S)-mandelic acids were -153° and +154°, respectively, in H_2O. Complexes with mandelic acid to monomer repeat unit ratios of 1:2, 1:1.5, 1:1 and 5:1 were prepared.

The ICD spectra in the solid state, see Figure 2, of P3M4VP/(R)- and (S)-MA complexes, prepared in THF at a MA to monomer repeat unit ratio of 1:2, show two mirror image Cotton effect signals at 211 and 214 nm, indicating formation of enantiomeric helical structures. Helicity selection or chaperoning to some higher enantiomeric structural order was observed for complexes prepared in THF, H_2O and CH_3OH at MA to monomer repeat unit ratios of 1:1, 1:1.5 and 1:2.

Is it possible to carry out such chaperoning and successfully observe the ICD spectra of the acid-base complexes in solution? Preliminary studies indicate that this is indeed possible. The ICD spectra of P3M4VP/(R)- and (S)-MA complexes (see figure 3) in tetrahydrofuran at a MA to monomer repeat unit ratio of 1:2, show mirror image Cotton effect signals at 280 nm, 283 nm and 287 nm, suggesting the formation of enantiomeric structures *(16,17)*. The CD spectroscopy indicates that the secondary structure of the P3M4VP/(R)- and (S)-MA complexes, once formed, is stable in room temperature. This is contrary to the observation of the uncomplexed (+) and (-) helical P3M4VP *(8)*. Detailed studies on the "chaperoning" of poly(3-methyl-4-vinylpyridine) by chiral acid-based interactions are currently in progress.

Conclusion

Helix sense selective polymerization of 3-methyl-4-vinylpyridine is possible and the resulting secondary helical conformation is stable in the solid state and in the solution at -78°C. Molecular modeling (MM3 and PM3) and conformational studies suggest that the secondary structure is a helix. The helical conformation of poly(3-methyl-4-vinylpyridine) may be easily locked in a solid matrix and the conformation in the matrix is stable at room temperature. Additionally, helicity selection of a single-screw sense starting with a racemic mixture of helical poly(3-methyl-4-vinylpyridine) may be carried out via chiral acid-base interactions.

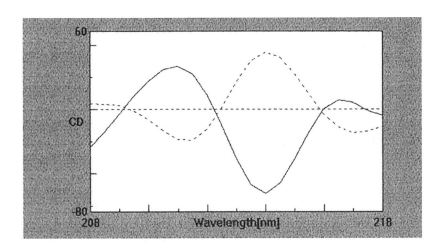

Figure 2: Solid state ICD (Induced circular dichroism) spectra of P3M4VP/(S)-MA (-----) and P3M4VP/(R)-MA (———) at an MA to monomer repeat unit of 1:2 prepared in THF at room temperature.

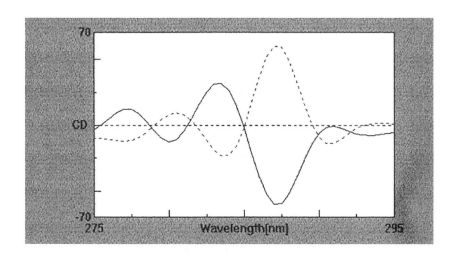

Figure 3: Solution ICD spectra of P3M4VP in the presence of (R)-MA (———) and (S)-MA (-------) in THF at room temperature.

64

Acknowledgement: Support of this work by NIH/NIGMS/MBRS/SCORE Grant #SO6GM08247, RCMI Grant #G12RR03062 and NIH/MIRT # 5T37TW0091 is gratefully acknowledged.

References and Notes

1. Fujiki, M. *J. Am. Chem. Soc.* **2000**, *122*, 3336-3343.
2. Okamoto, Y.; Nakano, T. *Chem. Rev.* **1994**, *94*, 349-372.
3. (a) Corley, L.S.; Vogl, O. *Polym. Bull.* **1980**, *3*, 211-217. (b) Kamer, P.C.J.; Nolte, R.J.M.; Drenth, W. *J. Am. Chem. Soc.* **1988**, *110*, 6818-6825. (c) Green, M.M.; Peterson, N.C.; Sato, T.; Teramoto, A.; Cook, R.; Lifson, S. *Science*, **1995**, *268*, 1860-1865. (d) Okamoto, Y.; Mohri, H.; Nakano, T.; Hatada, K. *J. Am. Chem. Soc.* **1989**, 111, 5952-5954. (e) Vogl, O., *J. Poly. Sci. Poly. Chem*, **2000**, 2623-2634.
4. (a) Okamoto, Y.; CHEMTECH, **1987**, 144. (b) Okamoto, Y.; Honda, S.; Okamoto, I.; Yuki, H.; *J. Am. Chem. Soc.* **1981**, *103*, 6971-6973.
5. Unpublished studies from our laboratory have shown that optically active helical (+)-poly(diphenyl-2-pyridylmethacrylate)/BH$_3$ complexes reduces acetophenone to (R)-(+)-sec-phenethyl alcohol in enantiomeric excesses as high as 65%. However, the process suffers from two drawbacks. First the chemical yield is low (7-8%) and this is most likely because while BH$_3$ may penetrate the swollen polymer and form the polymeric reagent, the larger acetophenone molecule may not do so, thus directly affecting the chemical yield. Secondly, while workup for isolation of the product is relatively easy, it is not possible to effectively regenerate the polymer reagent after one cycle. This is because the optical activity of the polymer decreases during the reduction process and after 20 hours the optical activity is negligible. Most likely this is due to the helix-helix interconversion resulting in racemization and/or side reactions that may be taking place during reduction. For details see, Yun Liu, "Asymmetric Reduction of Porchiral Ketone by Optically Active Polymeric Reagents", MS Thesis, Clark Atlanta University, **1995**.
6. Cornelissen, J.J.L.M.; Fischer, M.; Sommerdijk, N.A.J.M.; Nolte, R.J.M. *Science* **1998**, *280*,1427-1430.
7. Hirschberg, J.H.K.K.; Brunsveld, L.; Ramzi, A.; Vekemans, J.A.J.M.; Sijbesma, R.P.; Meijer, E.W. *Nature* **2000**, *407*, 167-170.
8. Ortiz, L.J.; Khan, I.M. *Macromolecules* **1998**, *31*, 5927-5929.
9. Habaue, S.; Ajiro, H.; Okamoto, Y. *J. Polym. Sci. Poly. Chem.* **2000**, *38*, 4088-4094.

10. (a) Bovey, F. A.; Jelinski, L. W., "Chain Structure and Conformation of Macromolecules", Academic, Press, **1992**. (b) Natta, G.; Danusson, F.; Sianesi, D. *Makromol. Chem.* **1958**, *28*, 253.

11. Khan, I. M.; Hogen-Esch, T. E. *Macromolecules* **1987**, *20*, 2335-2340.

12. The composites were prepared using PEO (MW 600,000) and PEG (MW 3000) to obtain a thermoplastic elastomeric material. Elastomeric materials are easy to fabricate for CD measurements. Temperature dependent racemization studies were not carried out but since segmental motion seems to be necessary for helix-to-helix interconversion, the racemization in the solid state should be possible above the glass transition temperature of the polymer matrix. Rate of racemization should increase with increasing temperature. Detailed DSC studies have not been undertaken, PEG is a good plasticizing agent, and additionally because PEO has a melting temperature around 60 °C, the crystalline content of the composite at this temperature is small.

13. Alchemy 32 Version 2.05, Tripos, Inc. St. Louis, MO, USA

14. (a) Maeda, K.; Yamamoto, N.; Okamoto, Y.; *Macromolecules* **1998**, *31*, 5924-5926. (b) Schlitzer, D.; Novak, B.M. *J. Am. Chem. Soc.*, **1998**, *120*, 2196-2197. (c) Yashima, E.; Matsushima, T.; Okamoto, Y. *J. Am. Chem. Soc.* **1995**, *117*, 11596-11957. (e) Yashima, E.; Nimura, T.; Matsushima, T.; Okamoto, Y. *J. Am. Chem. Soc.* **1996**, *118*, 9800-9801. (f) Yashima, E.; Matsushima, T.; Okamoto, Y. *J. Am. Chem. Soc*, **1997**, *119*, 6345-6359. (g) Yashima, E.; Maeda, K.; Yamanaka, T. *J. Am. Chem. Soc.*, **2000**, *122*, 7813.

15. The usage of the term "chaperoning" is appropriate in this system. Starting with a racemic helical poly(3-methyl-4-vinylpyridine), an optically active chaperone is utilized to carry out a helicity selection process.

16. The complexes were prepared by mixing 5 mg of P3M4VP and the appropriate amount of the MA in 5 ml of THF and stirred for 6 hour at room temperature.

17. Mirror image Cotton effect signals in the solid state have been observed for P3M4VP complexes with d- and l-alanine between 265 and 275 nm for samples prepared in CH_3OH/H_2O. In figure 2, the Cotton effect signals may due to the absorption of the carbonyl of MA.

Chapter 6

Helical Poly(diarylsilylenes): Effects of Higher Order Structure on Optical Activity

Julian R. Koe[1,2], Michiya Fujiki[1,2,*], Hiroshi Nakashima[1,2], and Masao Motonaga[2]

[1]NTT Basic Research Labs, and [2]CREST-JST, 3–1 Morinosato, Wakamiya, Atsugi, Kanagawa 243–0098, Japan

Poly(diarylsilylenes) with side chains bearing enantiopure chiral groups are observed to exhibit optical activity due to the induction of helical higher order structure resulting from the preferential stereochemical interaction between the side chains. The helical screw sense and degree of cooperative order depend on temperature, position of aryl ring substitution of the chiral group, number of chiral centers per monomeric unit and for copolymers on the chiral:achiral monomer ratio, allowing chemical and physical control of the sign and magnitude of optical activity.

While the origin of chirality, or more precisely, homochirality, is one of the most encompassing and hotly debated topics of science, attracting inter alia biologists, chemists, physicists, and astronomers, what is not in question is our dependence on homochirality for life as we know it (1-3). The fundamental building blocks of life are homochiral and thus optically active: those which make up proteins are composed of L-amino acids and our own DNA comprises D-sugars. In addition to being composed of homochiral units, proteins and DNA themselves forms chiral superstructures (4) - single screw sense helical macromolecules - and are thus optically active for two reasons. In contrast, many synthetic polymers, such as polyisocyanates, polyisocyanides, polyphenylenevinylenes, polyacetylenes, polythiophenes and polysilylenes, exist as racemates, without the higher structural order (and the concommitant bulk properties) imparted by a helical main chain conformation (5). However, several methods exist by which preferential screw sense helical polymers may be obtained (6, 7). These include: (a) polymerization with unichiral catalyst (8) or initiator (9), (b) chiral doping of achiral (racemic) polymers with enantiopure chiral ions (10-13), (c) separation of a racemic mixture of enantiomeric helices using chiral stationary phase (CSP) chromatography (for non-dynamic systems) (14), (d) chiral complexation of achiral or racemic polymers with enantiopure chiral ligands / guests (15-18), (e) post-polymerization functionalization with chiral moieties (19), (f) incorporation of enantiopure chiral end groups (20, 21), (g) polymerization of enantiopure chiral monomers (22-24) and (h) copolymerization of enantiopure chiral monomers with achiral monomers or with an enantiomeric excess (*ee*) of one enantiomer over the other (25-28). In the last two cases, it is the presence of the enantiopure chiral "seeds" which result in the adoption of a preferential helical screw sense in even non-enantiopure substituted backbone regions by the preferential stereorelationship between enantiopure chiral sidechains and their nearest neighbours. These cooperative phenomena (29) have been referred to as "sergeants and soldiers" or "majority rules" to describe the non-linear relation between the specific optical rotation and the *ee* of chiral units in helical polythiophenes (30) and polyisocyanates (31) and have also been observed in certain poly(dialkylsilylene) systems (25).

Optical activity may be analyzed by simple optical rotation, but much more comprehensive data can be obtained by circular dichroism (CD) spectroscopy (32) if the molecule to be investigated contains an optically active chromophore, as do polysilylenes. CD is essentially the same as ultraviolet (UV) spectroscopy, except that two separate light beams pass through the sample, one of right circularly polarized (CP) light and the other of left CP light. The intensity difference between left and right CP light absorption is called the circular dichroism, $\Delta\varepsilon = \varepsilon_L - \varepsilon_R$ (where ε is the molar absorptivity and the subscripts refer to the polarization of the light), which is output often simultaneously with the UV spectrum. The ratio of CD to UV absorption intensity is called the Kuhn dissymmetry ratio and is defined as $g_{abs} = \Delta\varepsilon/\varepsilon$ (33). This dimensionless quantity is perhaps the most useful in comparing the CD spectra of chromophoric molecules, since it provides a gauge of CD independent of UV dipole strength, and will be discussed for the poly(diarylsilylenes) in this study.

Polysilylenes (34, 35) are hybrid inorganic-organic polymers comprising a main chain of catenating tetravalent silicon atoms, each with two organic (usually alkyl or aryl) side chains. A number of excellent reviews (36-39) describe the fundamental properties and potential applications (including electroluminescence, EL) of these materials, which, due to σ-σ* conjugation of backbone silicon orbitals, are inherent semiconductors. The conformations of these polymers in both solution and solid state is a topic of brisk debate and complex, since side chain type, solvent, or lack thereof, and temperature are important factors. Simply put, for polysilylenes in solution, descriptions of morphology have been given on the microscopic scale as of helical or all anti (all trans) conformations, and on the macroscopic, as random coil or rod-like. One of the most sensitive and informative properties of polysilylenes is the UV absorption due to the σ-σ* transition noted above, which typically occurs in the range 300 - 400 nm. The energy of this transition markedly depends conformationally and electronically on the backbone dihedral angle and orbital mixing with side chain substituents (40-42). The longest wavelength absorptions are observed for poly(diarylsilylenes), the electroluminescent (EL) properties of which have been investigated in our labs, since among polysilylenes, they exhibit the best characteristics (43-45). The low transition energy of the diaryl-substituted polysilylenes was suggested to arise possibly from an all anti conformation, even in solution, although the red-shifting effect of substituent phenyl groups was also discussed (46).

As part of our investigations, we therefore undertook to determine whether poly(diarylsilylenes) exist in the all anti conformation in solution, or whether they are in fact helical, as represented in Scheme 1.

Scheme 1. Representations of (a) all anti and (b) helical conformations.

Our strategy was to synthesize a number of poly(diarylsilylene) homo- and copolymers bearing enantiopure chiral alkyl substituents on the phenyl rings, as in **1 - 13**, such that in the case that poly(diarylsilylenes) are helical, then preferential stereochemical interactions between the enantiopure chiral side chains on neighboring silicon atoms should induce a preference for one helical screw sense over the other, either *P* (plus) or *M* (minus). Since the silicon σ-σ* transition chromophore would then be present in a homochiral helical environment, a CD signal should be observed at the same wavelength as the UV absorption. We now describe our results, both experimental and theoretical, which clearly show that poly(diarylsilylenes) adopt helical backbone conformations, and further that the CD sign (related simply to helical screw sense) and intensity (a function of screw pitch, helical purity and dipole strength) are dependent on the chiral center distribution and type, and temperature. The magnitude and sign of the optical activity is thus controllable both chemically and physically. Some of these results have been previously published and are referenced: effect of chiral center number density on homopolymers (22), temperature-dependent helix-helix transition (47) and helical cooperativity in copolymers (48).

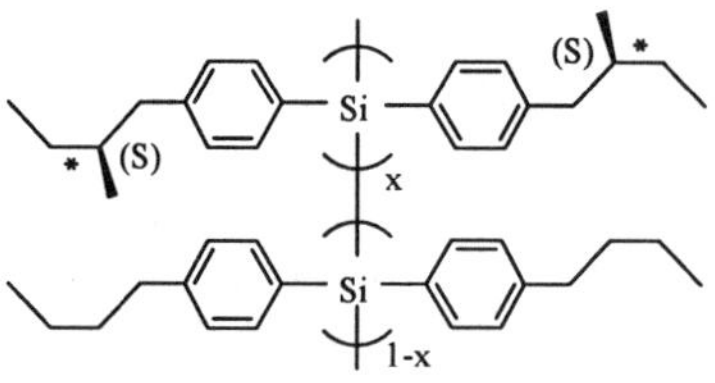

1: x = 0.0
2: x = 0.2
3: x = 0.5
4: x = 0.8
5: x = 1.0

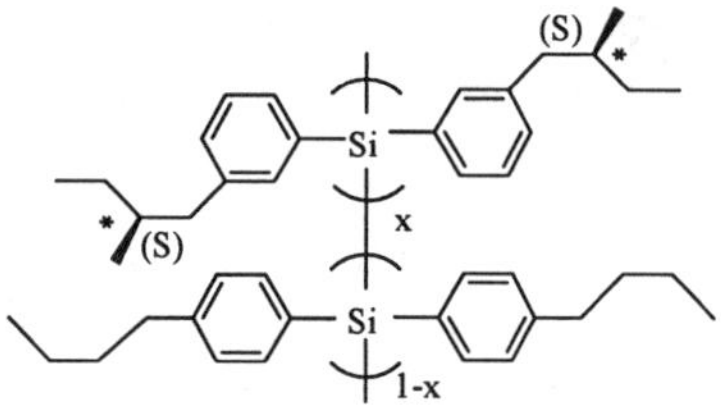

6: x = 0.2
7: x = 0.5
8: x = 0.8
9: x = 1.0

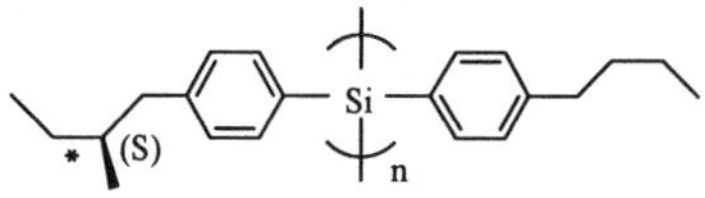

10

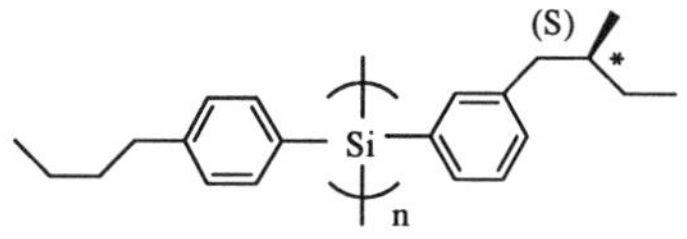

11

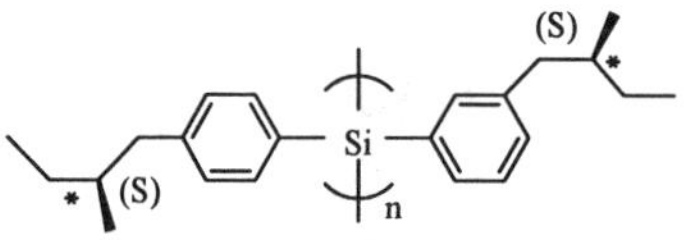

12

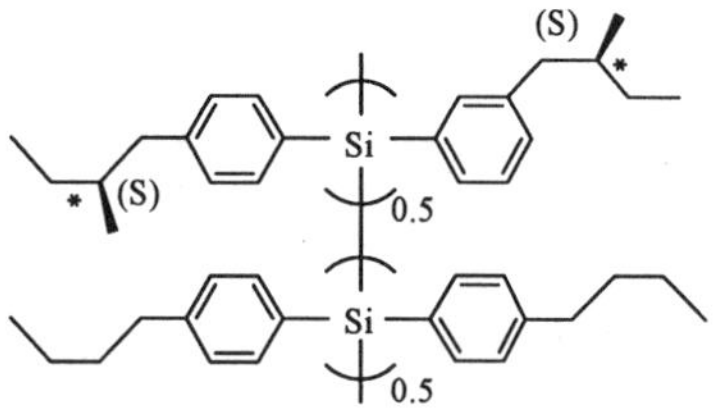

13

Experimental

Molecular weights were determined by size exclusion chromatography (SEC) on a Shodex column (eluant THF) in a Shimadzu liquid chromatograph machine equipped with a photodiode array detector and calibrated with polystyrene standards. UV-Vis spectra (room temperature; 21 °C) were recorded on a JASCO V-570 spectrometer at 1×10^{-4} mol dm^{-3} concentration. Variable temperature circular dichroism (CD) and simultaneous UV-Vis spectra were recorded using a JASCO J-720 spectropolarimeter and Peltier control for temperatures from 80 to -10 °C (1 cm path length cell; sample concentration 1×10^{-4} mol dm^{-3}), and with a liquid nitrogen-cooled cryostat for temperatures down to -70 °C (0.5 cm path length cell; sample concentration 2×10^{-4} mol dm^{-3}). Fluorescence spectra were recorded on a Hitachi F-850 spectrofluorimeter at room temperature. NMR spectra were recorded on a Varian Unity 300 spectrometer relative to internal TMS in toluene-d$_8$ for silicon at 59.591 MHz and for carbon at 75.431 MHz. Enantiopure chiral *para-* and *meta*-substituted (*S*)-2-methylbutylphenylbromides were custom synthesized by Chemical Soft Co., Kyoto, Japan, and purified prior to use by distillation using a Perkin Elmer autoannular still.

Monomers

For the symmetrical monomers, these were prepared by lithiation of the respective arylbromides in ether at 0 °C using *n*-butyl lithium and addition of the resulting aryl lithium solution to 0.5 equivalents of silicon tetrachloride in hexane at 0 °C. For the assymmetrical monomers, a trichlorosilane was initially prepared from one aryl lithium compound and silicon tetrachloride, similar to the above and purified, and to this was added the second aryl lithium. Filtration of the solution and distillation under reduced pressure afforded > 95% pure (by GC) monomers in *ca.* 60% yield.

Polymers

Poly(diarylsilylenes) **1 - 13** were prepared according to the Wurtz-type reductive coupling method, for homopolymers using a single monomer, and for copolymers using a mixture of the two monomers. Since this method involves the reaction of sodium metal with reactive chlorosilanes in flammable solvents, care should be taken to control the temperature and eliminate air and water, as otherwise there is potential for violent reaction or ignition. The synthesis of poly[{bis-*p*-(*S*)-2-methylbutylphenyl}$_{0.5}$-*co*-(bis-*p*-*n*-butylphenyl)$_{0.5}$silylene], (**7**) is representative: Bis{*p-(S)*-2-methylbutylphenyl}dichlorosilane (1.52 g, 3.86 mmol) and bis(*p-n*-butylphenyl)dichlorosilane (1.41 g, 3.86 mmol) were combined to give a mixture in the ratio 50:50 and added in toluene (2 mL) to 2.5 equivs. sodium (0.44 g, 19.32 mmol) dispersed in toluene (11 mL) and surface-activated by diglyme (13 µL), under an argon atmosphere. The mixture was stirred slowly at 70°C, shielding the vessel from light and monitoring the molecular weight periodically by SEC. After 3 hours, 0.1 equivs. trimethylchlorosilane (0.08 g, 0.77 mmol) was added to terminate the reaction and stirring continued for a further 30 minutes, after which the mixture was filtered through two Teflon membranes (pore size 40 µm and 10 µm) under pressure of argon gas. The high molecular weight (M_w) fraction was isolated as a white powder or fibrous material by fractional precipitation in a mixed isopropanol-toluene solution, followed by centrifugation, and vacuum dried in an oven at 80°C overnight. Yield: 0.24 g, (10.1 %). Incorporation of the monomers at the nominal addition ratio was verified by ^{1}H and ^{13}C NMR spectroscopy.

Viscometric Studies

Measurements were performed by the Toray Research Center (Shiga, Japan) using an in-line configuration of viscometer (Viscotek H502a, equipped with a capillary of dimensions 0.5 mm i.d. × 61 cm length) and SEC (Waters 150C). The solvent was THF and measurement temperature 30°C. Molecular weights were determined from a Universal Calibration plot and the corresponding intrinsic visosities, [η], were obtained from the viscometer. Application of the Mark-Houwink-Sakurada equation, [η] = KM^a, in a plot of log[η] vs. log molecular weight, M, affords the viscosity index a, and constant K, from the slope and intercept, respectively. The radius of gyration, R_g, at a given molecular weight, M, may be derived from [η]$M = \Phi(6R_g^2)^{3/2}$, where Φ is the Flory constant (2.86 × 10^{-23} mol^{-1}).

Results and Discussion

Data for poly(diarylsilylenes) **1 - 13** are given in Table 1. The Wurtz-type reductive coupling of diaryldichlorosilane monomers at 70°C yielded soluble high molecular weight (M_w) poly(diarylsilylenes) in approximately 2 - 10% isolated yield as white fibrous or powdery materials.

Dissolved in isooctane or toluene, poly(diarylsilylenes) **2 - 13** containing enantiopure chiral substituents show relatively narrow ($\Delta v_{1/2}$ ca. 15 - 20 nm) UV absorptions (peak maximum ca. 396 nm) due to the conjugated Si backbone σ-σ^* transition, mirror image fluorescence emission spectra, small Stokes' shifts (ca. 13 nm) and fluorescence anisotropies in the range 0.20 - 0.38, indicating regular, semi-flexible polymer molecules with long segment lengths (22). Circular dichroism spectra evidenced a Cotton effect at the same wavelength as the UV absorption, such optical activity arising from the adoption of a preferential screw sense helical backbone conformation. The CD characteristics for the various polymers, however, are very different, depending on side chain chiral center distribution and type (ie. whether homopolymer or copolymer, aryl ring substitution position and whether symmetrical or otherwise), and temperature.

Circular Dichroism and Ultraviolet Spectroscopy

Homopolymers

The 20°C CD spectrum of **9** (Figure 1a), the homopolymer containing 100% bis-m-(S)-2-methylbutylphenyl-substituted monomer units, exhibits a negatively signed Cotton effect (385 nm), almost coincident with the lowest energy backbone electronic transition (σ-σ^*) in the 20°C UV spectrum (388 nm). This CD signal derives from the preferential absorption of right circularly

Table I. Data[a] for Polymers 1 - 13

Cpd	T /°C	UV ε/λ$_{max}$/fwhm	CD Δε/λ$_{ext}$	$10^{-4}g_{abs}$	FL λ$_{max}$	FL-A[b]	M_w /10^3	PDI	Yld %
1[d]	20	13,300 / 394 / 15.8	0.00 / -	0.00	407.0	0.20-0.35	434	2.9	2.5
	-10	15,410 / 392 / 13.8	0.00 / -	0.00					
2	20	10,900 / 393 / 16.8	0.45 / 391	0.41	406.5	0.25-0.35	162	3.2	1.6
	-10	12,300 / 392 / 15.1	1.04 / 388	0.84					
3	20	17,400 / 394 / 15.0	1.73 / 389	0.99	407.0	0.27-0.33	186	2.7	10.0
	-10	19,300 / 392 / 13.6	3.67 / 387	1.90					
4	20	14,700 / 395 / 17.1	0.81 / 394	0.55	407.0	0.18-0.28	247	3.3	6.9
	-10	14,400 / 394 / 16.0	1.63 / 390	1.13					
5[d]	20	8,400 / 395 / 18.7	-1.47 / 384	-0.35	406.0	0.20-0.30	240	2.8	3.2
6[c]	20	24,300 / 397 / 14.0	0.24 / 393	0.10	408.5	0.20-0.30	474	2.8	5.9
	-10	26,800 / 396 / 12.9	0.04 / 390	0.02					
7[c]	20	25,700 / 399 / 14.6	-0.61 / 400	-0.24	410.5	0.20-0.30	271	2.9	3.9
	-10	28,200 / 398 / 13.2	-0.65 / 397	-0.23					
8[c]	20	10,400 / 393 / 19.0	-1.06 / 392	-1.02	407.5	0.34-0.38	8	1.3	3.3
	-10	11,100 / 392 / 17.5	-1.37 / 389	-1.23					
9[d]	20	7,400 / 386 / 26.5	-1.60 / 382	-2.16	404.0	0.30-0.38	5	1.1	3.3
	-10	7,800 / 384 / 23.4	-1.86 / 384	-2.38					
10	20	10,500 / 396 / 19.3	0.50 / 395	0.48	407.5	0.20-0.30	333	2.8	3.3
11	20	7,900 / 397 / 21.7	0.27 / 396	0.34	411.0	0.20-0.30	254	2.6	7.4
12	20	11,100 / 399 / 18.7	0.16 / 398	0.14	409.0	0.27-0.39	514	3.5	11.2
13	20	19,400 / 398 / 15.0	-0.19 / 398	-0.10					4.8

a UV and CD data in isooctane at -10 and 20°C; fluorescence (FL) data in toluene at room temperature (21°C). ε and Δε units: (Si repeat unit)$^{-1}$dm^{3}cm^{-1}; λ$_{max}$ and λ$_{ext}$ units: nm; fwhm = full width (nm) at half maximum of λ$_{max}$. Size exclusion chromatography (SEC) data: molecular weights determined by SEC and relative to polystyrene standards; eluant: THF; PDI = polydispersity index: M_w/M_n; given for isolated high M_w fractions; yields given for isolated high M_w fraction.
b fluorescence anisotropy measured over region of backbone σ-σ* transition; toluene, room temperature.
c data taken from Ref. 47.
d samples described in Ref. 22; data newly recorded (in isooctane for 1 and 9 and in toluene for 5 due to poor solubility in isooctane). For 5, see also note in ref. 49.

polarized light and indicates that **9** adopts a preferential screw sense helical backbone conformation, which presumably results from the preferential stereorelationship between phenyl ring alkyl substituents on neighboring silicon atoms.

The Kuhn dissymmetry ratio for **9** at 20°C is -2.16×10^{-4}. It has been suggested that in the case of poly[{*(S)*-2-methylbutyl}(6,9,12-trioxatetradecyl)-silylene], a g_{abs} value of 2.8×10^{-4} (at -60°C) may correspond to a single screw sense helical polysilylene main chain (50). It is likely, therefore, that the catenating Si backbone of **9**, while helical, contains pseudoenantiomeric helical segments (ie. both plus (*P*) and minus (*M*) screw senses), one of which predominates. In contrast to that of **9**, the CD spectrum of **11**, the assymmetric homopolymer bearing one *m*-(S)-2-methylbutylphenyl and one *p*-*n*-butylphenyl group per silicon atom, exhibits a positive Cotton effect (396 nm) at the lowest energy UV transition, indicating that the backbone adopts the opposite preferential helical screw sense to that for **9**, although the small Kuhn dissymmetry ratio suggests the preference is small.

Similarly, the CD spectrum of **10** (Figure 1b), the *para* analog of **11**, shows a positive Cotton effect (395 nm), though the g_{abs} value is greater. It could be argued that a small g_{abs} value, as in the case of **11**, is the result of an almost planar all-anti backbone conformation; however, an all-anti backbone would be expected to have a significantly red-shifted absorption due to the dependence of σ-σ* transition energy on dihedral angle (40). Since the CD extrema are almost the same, it is reasonable to suggest that the small g_{abs} value stems rather from a backbone of similar screw pitch, but weaker helical screw sense preference, such that partial cancellation of oppositely signed CD bands of major and opposite minor screw sense segments occurs. The weaker screw sense preference in the case of **11** may be a result of the greater ability to mitigate steric congestion by phenyl ring rotation to minimise interaction between the *meta* chiral alkyl groups on nearest and next nearest silicon atoms.

One other possible factor affecting CD and UV data is the helical domain size (polymer chain length divided by number of helix reversals, the latter possibly reducing conjugation): where the helical domain size is larger than the effective conjugation length of the silicon backbone, there should be no significant effect on the UV energy or intensity, although the resultant CD will depend on the position of the helix reversal. Where, however, steric constraints impose a helical domain size shorter than the effective conjugation length, the UV peak should undergo a blue shift and possibly broaden. The CD wavelength would also undergo a blue shift, though the sign of the effect should depend rather more on the relative proportions of *P* and *M* screw sense domains, rather than the size of the domain. Unambiguous experimental support for these hypotheses, however, is still required.

The symmetrical *para* homopolymer, **5**, exhibits an essentially negative CD band at 20°C, although as we noted previously, a small positive band is evident on the longer wavelength side, which we attributed to the minor presence of longer pitch helical segments, opposite in screw sense to the major, shorter pitch segements and due to the pitch difference, not cancelling out (pseudo-diastereomeric helices).

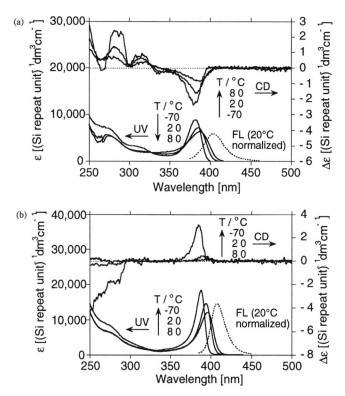

*Figure 1. CD, UV and Fluorescence spectra of (a) **9** and (b) **10**. (Adapted with permission from reference 22. Copyright, ACS 1999.)*

Further experiments confirm our belief that the origin of the negative and positive CD bands in **5** is the coexistence of both *P* and *M* screw senses, rather than a bisigned CD exciton couplet (Davidov) effect (51), which is observed for chromophores interacting intramolecularly (eg. two segments either side of a strong kink, or phenyl rings) or intermolecularly (eg. aggregation effects). This is because spectra exhibiting the Davidov phenomenon are very different, the bands being usually of significantly greater intensity than those for molecularly disperse polymers. An example of aggregation effects is given in Figure 2, which shows a bisigned CD exciton couplet arising from aggregation of **10** induced by addition of a poor solvent (methanol) to a toluene solution of the polymer. The bisigned Cotton effect (conventionally taking the sign of the longest wavelength extremum; ie. in this case, positive) is relatively intense, ($\Delta\varepsilon$ = 3 and -5 (Si repeat unit)$^{-1}$dm^{3}cm^{-1}). The only symmetrical homopolymer to show only positive CD is **12**, for which both rather low intensity CD and UV signals are observed. The origin of the former is likely to be in the coexistence of both helical screw senses, as described above, and the latter possibly in the existence of a large number of conjugation-limiting helix reversals, such that the conjugating segment lengths are limited by helical domain size.

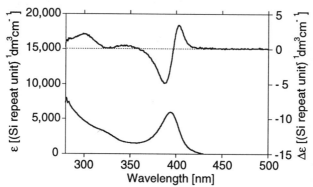

*Figure 2. Coupled excitonic CD for **10** due to aggregation in mixed solvent system (20 °C; MeOH:toluene = 40:60; 6×10^{-5}moldm^{-3}).*

It is thus clear that incorporation of enantiopure chiral side chains in poly(diarylsilylene) homopolymers affords optically active materials through the induction of helical higher order. Further, using only the *(S)* chiral form (cheaper and more readily available from natural products) of the 2-methylbutyl group, both *P* and *M* screw sense polymers may be prepared, depending on the number and aryl ring position of the chiral centers. Thus far we have demonstrated chemical control of the sign of optical activity through substituent effects on polymer higher order helical screw sense.

Copolymers

Whereas the CD spectrum of homopolymer **5** exhibits both negative and positive Cotton effects, the CD spectra of the related copolymers, **2 - 4**, are clearly positive, and, interestingly, of maximum intensity for **3** at -70°C ($\Delta\varepsilon$ = 13.9 (Si repeat unit)$^{-1}$dm^{3}cm^{-1}), with a 1:1 ratio of bis-*p*-*(S)*-2-methylbutylphenylsilylene to bis-*p*-*n*-butylphenylsilylene units. Since spectroscopic data (UV, fluorescence, fluorescence excitation (52) and fluorescence anisotropy) are similar for all the polymers (see Table 1), it is reasonable to assume that polymer backbone dihedral angles are also similar (since the UV absorption wavelength is very sensitive to the Si-Si-Si-Si dihedral angle), so that an increase in g_{abs} should thus imply greater helical screw sense selectivity. The larger g_{abs} values for **3** compared with **2**, **4** and **5** may therefore indicate that this composition corresponds to the maximum ratio of helical directing power to steric overcrowding (which we suggest may be responsible for the induction of helical reversal states).

Figure 3 shows a plot of dissymmetry ratio as a function of chiral content for the copolymers **2** - **4** (*para*) and **6** - **8** (*meta*) and their respective homopolymers **5** and **9**, together with data for a homochiral/achiral poly(dialkylsilylene) copolymer series, poly{n-hexyl[(S)-2-methylbutyl]$_x$-co-[n-hexyl(2-methylpropyl)]]$_{1-x}$}silylene, Si*Si (**25**), the latter exhibiting a "positive" cooperative effect.

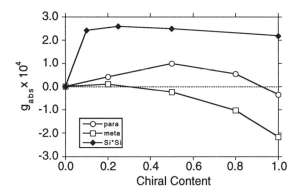

*Figure 3. Dependence of dissymmetry ratio on chiral content (para and meta copolymers at 20°C; Si*Si (data from ref. 25) at -5°C).*

A different trend is evident in the *meta* case: the dissymmetry ratio increases non-linearly with increasing chiral content, the maximum being for the homopolymer ("negative" cooperativity). Comparing the CD spectra of the 1:1 chiral:achiral *para* and *meta* copolymers, **3** and **7**, the magnitude of the Cotton effect is approximately a factor of 4 smaller for the latter (dissymmetry ratio at 20°C, g_{abs} = -0.24 × 10^{-4}), and the sign opposite, indicating that there is only a weak cooperative effect in the *meta* case, the backbone adopting the opposite screw sense to the *para*-substituted cases, presumably resulting from the better alleviation of steric hindrance afforded. Two possible reasons can be considered to account for the lower dissymmetry ratio: (i) that the polymer backbone dihedral angles are very close to 180°, resulting in a looser helix, and (ii) that although the backbone is helical, there is less dominant screw sense selectivity, possibly as a result of the greater ability of **7** to minimise substituent interactions between the *meta* chiral alkyl group and p-n-butyl or m-(S)-2-methylbutyl groups on neighboring silicon atoms by phenyl ring rotation, as discussed for homopolymer **11**, above, thus giving rise to weaker Cotton effects through partial cancellation of oppositely signed CD bands.

As can be seen by comparing the data in Table 1, the room temperature spectroscopic data (excluding CD) for all the polymers are similar, though **7** exhibits a longer UV absorption maximum [indicative of more open backbone dihedral angles (40, 41)] than **3**. We propose that the weaker CD signal for **11**, while deriving partly from a highly extended (near to all anti) backbone structure, is also the consequence of the coexistence, but in unequal proportions, of P and M sense helical turns.

The trends for both *para* and *meta* copolymers contrast with those seen in other optically active copolymer systems such as some polyisocyanates (53, 54) and poly(dialkylsilylenes) (25), for which the screw sense is determined by a small minority of enantiopure chiral repeat units and the dissymmetric ratio is approximately constant at greater chiral content levels. For these latter polymers, the non-linear relationship between optical activity and chiral content has been referred to as "sergeants and soldiers", the sergeants being the directing, homochiral monomer units and the soldiers the directed, achiral units. An example of a "sergeants and soldiers"-type response is afforded by the copolymer series, poly{n-hexyl[(S)-2-methylbutyl]$_x$-co-[n-hexyl(2-methylpropyl)]$_{1-x}$silylene (Si*Si), data for which are taken from ref. 25 and replotted in Figure 3.

Here we have demonstrated chemical control of the magnitude of optical activity, through cooperativity in higher structural order.

Force Field Calculation for Model Poly(diarylsilylene)

In order to test our experimental observations of helicity theoretically, we performed a series of force field calculations for the model poly(diarylsilylene), H-[(p-n-BuPh)$_2$Si]$_{30}$-H, using Discover 3 (ver. 4, standard pcff parameters, Molecular Simulation, Inc.). Results for a range of dihedral angles are plotted in Figure 4. A monomer unit was polymerized to a chain length of 30 silicon atoms. The backbone dihedral angle was then input and the structure allowed to develop at zero Kelvin to find the local minimum energy. While the choice of initial ground state structure affects the relative minimum energies of the dihedral angle potential curves, the most important feature remains unchanged: that the energy at a dihedral angle of 180° (all anti, or all trans, as it has been commonly termed) is not a minimum. In fact, minima are evident for dihedral angles pairs of 165/195° and 150/210° (P/M screw sense helical conformations, respectively), corresponding approximately to 15_7 and 7_3 helical conformations of either screw sense, respectively. We therefore suggest that all poly(diarylsilylenes) are in fact helical, even if the dihedral angles are small, though unless enantiopure chiral units are incorporated, preferential screw sense helical higher order is not evident and the polymer molecules are internal racemates. In view of the developing consensus within the polysilylene community with respect to definitions of conformations, the descriptor "transoid" for these polymers seems appropriate, given the presumed closeness of the dihedral angle to 180°.

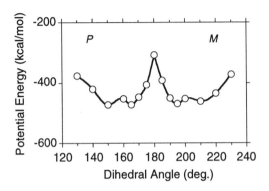

Figure 4. Force field-calculated potential energy as function of backbone dihedral angle for H-[(p-n-BuPh)₂Si]₃₀-H. (Reproduced with permission from reference 48. Copyright ACS 2000.)

Temperature-dependency of Helical Higher Order

In all the cases studied, the dissymmetry ratios are greatest at the lowest temperature, -70°C, and tend to zero at 80°C or above. This is consistent with the loss of helical screw sense preference in some segments (possibly end chain segments, which are less constrained, and segments next to them) due to the thermal energy exceeding that of the conformational fixing effect and consequent cancellation of oppositely signed CD bands. The majority of the polymers exhibit similar temperature dependency to that evident in Figure 1. Two polymers, however, depart notably from the general expectation: **6** in undergoing a rare example of a thermally driven helix-helix transition, and **4** in forming aggregates at low temperature with extremely intense Cotton effects.

Helix-Helix Transition

In the case of **6**, as we communicated recently (47), containing 20 % *m*-(S)-2-methylbutylphenyl-substituted monomer units, the CD spectrum shows a negative Cotton effect at -70°C, approximately zero intensity at -10°C and positively signed Cotton effects at temperatures above this, with a maximum at

50°C (see Figure 5), exhibiting a rare example of a temperature-driven helical inversion of a polymer in dilute solution.

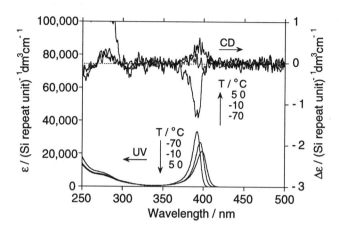

Figure 5. Temperature-dependency of CD and UV spectra of 6.
(Reproduced with permission from reference 47.
Copyright The Royal Society of Chemistry 2000.)

There are only five other reports of such inversions: one concerning certain poly(α-amino acid ester)s (55), another poly(β-phenethyl *L*-aspartate) (56), the third, a polyarylisocyanate (57), the fourth, poly(dialkylsilylenes) (58) and most recently, some polyisocyanate copolymers (59). The change of sign of the Cotton effect evident above indicates a change in the backbone conformation from one prevailing helical screw sense to the other. In the present case, since we consider that *P* and *M* screw senses coexist in these polymer systems, the transition may be considered to reflect the gradual thermodynamic stabilization of one screw sense relative to the other, as the populations alter with temperature, rather than an abrupt switch. The significance of this helix-helix transition is that it comprises *physical* control over the sign and magnitude of optical activity.

Low Temperature Aggregation

While **4** exhibits similar temperature-dependency to the others at temperatures down to -10°C in toluene (positive Cotton effect), a dip in the center of the Cotton effect in the CD spectrum at -40°C is apparent, even though the effect is still positive. At -70°C, however, a large negative extremum ($\Delta\varepsilon = -10$ (Si repeat unit)$^{-1}$dm^{3}cm^{-1}) is evident in the middle of the remaining positive effect which we assumed to be due to aggregation from the similarity of the spectrum to that in Figure 2 and our results concerning aggregation of poly(arylalkylsilylenes) (60). In an attempt to prevent the polymer from aggregating, we changed solvents to THF, but the result was the opposite: an extremely intense bisgned positive exciton couplet Cotton effect (positve $\Delta\varepsilon$ = 190, negative $\Delta\varepsilon$ = -160 (Si repeat unit)$^{-1}$dm^{3}cm^{-1}) due to aggregation, as shown in Figure 6. The UV decreases at -40 and -70°C due to the effective decrease of solution concentration upon aggregation.

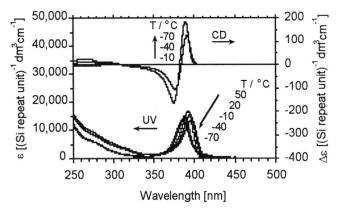

Figure 6. Intense bisignate exciton coupled CD due to aggregation of 4 in THF.

Viscometric Properties

In addition to showing that helical higher order structure results in bulk optical activity, we wished to investigate the macroscopic effect of incorporating enantiopure chiral monomer units of greater steric bulk than bis-*p-n*-butylphenylsilylene (by copolymerization) upon viscosity. To this end, the viscometric properties of **6** and **1** (data for **1** already reported (61), but measured here again for control purposes) were determined. The viscosity index, *a*, for **6**, **1** and reported **1** are the same (ca. 1.0 - 1.1) within experimental error (and variation expected due to differing experimental conditions between those described here and those in ref. (61). The major difference, however, lies in the data for the radius of gyration (R_g) of **6** and **1**, which are plotted in Figure 7 on a log/log scale, as a function of molecular weight [together with data for their idealized rigid rod analogs (62)] and are greater by about 30% for **6** compared to **1**. This is consistent with a stiffer, more extended structure of the copolymer **6** due to the greater steric demand of the bulkier branched alkyl phenyl side chains incorporated compared to **1**, the side chains of which contain straight chain alkyl phenyl ring substituents.

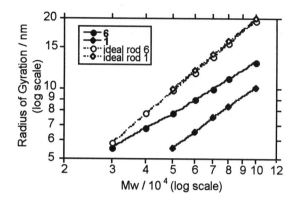

Figure 7. Radius of gyration, R_g, as a function of M_w for 6 and 1 in THF at 30 °C and their idealized analogs. (Adapted with permission from reference 48. Copyright ACS 2000.)

The R_g values for **6** also lie closer to those for the idealized rigid rod plot than do those of **1**. From the data represented in Figure 7, a greater persistence length for **6** compared to **1** could also be expected, and is indeed found (75 cf. 45 Å, respectively; estimates are derived in ref. 62). This result would also be consistent with the proposal that polysilylene UV absorption intensities are a function of viscosity index (63) and thus also of persistence length (as the latter two parameters are related (64, 65) through their dependence on R_g), since **6** has a much greater UV molar absorption coefficient (ε) than **7** (24,300 cf. 13,300 (Si repeat unit)$^{-1}$dm^{3}cm^{-1} at 20°C).

Conclusions

In summary, we have (i) succeeded in the preparation of the first optically active poly(diarylsilylenes), hybrid organic-inorganic chromophoric and fluorophoric polymers, exhibiting preferential screw sense helical higher order in their backbone conformations in solution, and chemical and physical control of the optical activities thereof. (ii) We have shown that the direction of the screw sense (and thus sign of the CD Cotton effect), both positive and negative, of helical poly(diarylsilylenes) containing (S)-2-methylbutylphenyl moieties is controllable by the number of chiral moieties per Si repeat unit. This side chain - main chain relationship should provide a new insight into the design of CP devices using only single-handed enantiopure chiral moieties. We have demonstrated (iii) cooperative helical ordering in poly(diarylsilylene) copolymers containing chiral and achiral monomer units. Dissymmetry ratios depend on temperature and the chiral:achiral monomer ratio and for the *para*-phenyl-substituted series, are greater than those of the related *meta*-phenyl-substituted copolymers and the homopolymers, indicating greater screw sense selectivity in the copolymers. This should also be useful knowledge in CP technology, as it significantly reduces the need for expensive enantiopure chiral synthons, while still affording optically active materials. (iv) We have demonstrated a structure-specific, reversible, temperature-dependent helical screw sense inversion for a copolymer, **6**. (v) Based on a comparison of spectroscopic data of polymers with chiral and achiral side chains and the results of force field calculations, we further suggest that poly[bis(*p-n*-butylphenyl)silylene] also adopts a helical backbone conformation comprising equal *P* and *M* sense turns in solution (hence its optical inactivity), and that poly(diarylsilylenes) generally should be considered to adopt transoid, rather than all-anti, conformations in solution. (vi) We finally show that incorporation of branched side chain monomer units results in a macroscopically stiffer, more extended polymer chain and provide further evidence for the dependence of polysilylene UV absorption intensities on viscosity index and persistence length.

References

1. Mason, S. F. In *Circular Dichroism: Principles and Applications*; Nakanishi, K.; Berova, N.; Woody, R. W., Eds.; VCH: New York, 1994; Chapter 2, p 39.
2. Feringa, B. L.; van Delden, R. A. *Angew. Chem. Int Ed.* **1999**, *38*, 3418.
3. Bailey, J.; Chrysostomou, A.; Hough, J. H.; Gledhill, T. M.; McCall, A.; Clark, S.; Menard, F.; Tamura, M. *Science (Washington D.C.)* **1998**, *281*, 672.
4. Petsko, G. A. *Science (Washington D.C.)* **1992**, *256*, 1403.
5. Schlitzer, D. S.; Novak, B. M. *J. Am. Chem. Soc.* **1998**, *120*, 2198.
6. Pu, L. *Acta Polymer* **1997**, *48*, 116.
7. Green, M. M.; Peterson, N. C.; Sato, T.; Teramoto, A.; Cook, R.; Lifson, S. *Science (Washington D.C.)*, **1995**, *268*, 1860.
8. Deming, T. J.; Novak,B. J. *J. Am. Chem. Soc.* **1992**, *114*, 7926.
9. Takei, F.; Koichi, Y.; Onitsuka K.; Takahashi, S. *Angew. Chem., Int. Ed. Engl.* **1996**, *35*, 1554.
10. Majidi, M. R.; Kane-Maguire, L. A. P.; Wallace, G. G. *Polymer* **1994**, *35*, 3113.
11. Majidi, M. R.; Kane-Maguire, L. A. P.; Wallace, G. G. *Polymer* **1995**, *36*, 3597.
12. Majidi, M. R.; Kane-Maguire, L. A. P.; Wallace, G. G. *Polymer* **1996**, *37*, 359.

84

13. Havinga, E. E.; Bouman, M. M.; Meijer, E. W.; Pomp, A.; Simenon, M. M. J. *Synth. Met.* **1994**, *66*, 93.
14. Nolte, R. J. M.; van Beijnen, A. J. M.; Drenth,W. *J. Am. Chem. Soc.* **1974**, *96*, 5932.
15. Yashima, E.; Matsushima, T.; Okamoto, Y. *J. Am. Chem. Soc.* **1995**, *117*, 11597.
16. E. Yashima, E.; Huang, S.; Matsushima, T.; Okamoto, Y. *Macromolecules* **1995**, *28*, 4184.
17. Yashima, E.; Maeda, Y.; Okamoto, Y. *J. Am. Chem. Soc.* **1998**, *120*, 8895.
18. Yashima, E.; Maeda, Y.; Okamoto, Y. *Nature (London)* **1999**, *399*, 449.
19. Terunuma, D.; Nagumo, K.; Kamata, N.; Matsuoka, K.; Kuzuhara, H. *Chem. Lett.* **1998**, 681.
20. Obata, K.; Kabuto, C.; Kira, M. *J. Am. Chem. Soc.* **1997**, *119*, 11345.
21. Obata, K.; Kira, M. *Macromolecules* **1998**, *31*, 4666.
22. Koe, J. R.; Fujiki, M.; Nakashima, H. *J. Am. Chem. Soc.* **1999**, *121*, 7934.
23. Fujiki, M. *J. Am. Chem. Soc.* **1994**, *116*, 6017.
24. Shinohara, K.; Aoki, T.; Kaneko, T.; Oikawa, E. *Chem. Lett.* **1997**, *6*, 361.
25. Fujiki, M. *Polym. Prepr. (ACS Div. Polym. Chem.)* **1996**, *37(2)*, 454.
26. Frey, H.; Möller, M.; Matyjaszewski, K. *Macromolecules* **1994**, *27*, 1814.
27. Frey, H.; Möller, M.; Turetskii, A.; Lots, B.; Matyjaszewski, K. *Macromolecules* **1995**, *28*, 5498.
28. Jha, S. K.; Cheon, K-S.; Green, M. M.; Selinger, J. V. *J. Am. Chem. Soc.* **1999**, *121*, 1665.
29. Ciardelli, F.; Chiellini, E.; Carlini, C. In *Optically Active Polymers*; Selegny, E., Ed.; D. Riedel Publishing Co.: Dordrecht, 1979.
30. Langeveld-Voss, B. M. W.; Waterval, R. J. M.; Janssen, R. A. J.; Meijer, E. W. *Macromolecules* **1999**, *32*, 227.
31. Green, M. M.; Reidy, M. P.; Johnson, R. J.; Darling, G.; O'Leary, D. J. Willson, G. *J. Am. Chem. Soc.* **1989**, *111*, 6452.
32. See *Circular Dichroism: Principles and Applications*; Nakanishi, K.; Berova, N.; Woody, R. W., Eds.; VCH: New York, 1994.
33. Dekkers, H. P. J. M. In *Circular Dichroism: Principles and Applications*; Nakanishi, K.; Berova, N.; Woody, R. W., Eds.; VCH: New York, 1994; Chapter 6, p 122.
34. Trujillo, R. E. *J. Organomet. Chem.* **1980**, *198*, C27.
35. West, R.; David, L. D.; Djurovitch, P. I.; Stearly, K. L.; Srinivasan, K. S. V.; Yu, H. J. *J. Am. Chem. Soc.* **1981**, *103*, 7352.
36. West, R. In *The Chemistry of Organosilicon Compounds*; Patai,S.; Rappoport, Z., Eds.; John Wiley and Sons: Chichester, 1989; Part II, Chapter 19, p 1207.
37. Miller, R. D.; Michl, J. *Chem. Rev.* **1989**, *89*, 1359.
38. West, R. *J. Organomet. Chem.* **1986**, *300*, 327.
39. Zeigler, J. M. *Synth. Met.* **1989**, 28, C581.
40. Michl, J. *Synth. Met.* **1992**, *49-50*, 367.

41. Takeda, K.; Teramae, H.; Matsumoto, N. *J. Am. Chem. Soc.* **1986**, *108*, 8186.
42. Harrah, L. A.; Zeigler, J.M. *Macromolecules* **1987**, *20*, 601.
43. Yuan, C.-H.; Hoshino, S.; Toyoda, S; Suzuki, H.; Fujiki, M.; Matsumoto, N. *Appl. Phys. Lett.* **1997**, *71*, 3326.
44. Suzuki, H; Hoshino, S.; Yuan, C.-H.; Fujiki, M.; Toyoda S; Matsumoto, N. *IEEE J. Sel. Top. Quant. Electr.* **1998**, *4*, 129.
45. Suzuki, H.; Yuan, C.-H.; Furukawa, K.; Matsumoto, N. *Polym. Prepr. (ACS Div. Polym. Chem.)* **1998**, *39(2)*, 996.
46. Miller, R. D.; Sooriyakumaran, R. *J. Polym. Sci.: Polym. Lett. Ed.* **1987**, *25*, 321.
47. Koe, J. R.; Fujiki, M.; Motonaga, M.; Nakashima, H. *Chem. Commun.* **2000**, 389.
48. Koe, J. R.; Fujiki, M.; Motonaga, M.; Nakashima, H *Macromolecules* submitted.
49. As noted in ref. 20 of our previous work (22), at 20°C, the CD band of **8** approximately coincident with the UV absorption at 395 nm is essentially negative, although a small, positive, longer wavelength band was apparent. At lower temperature, a new band in the UV spectrum was noted at about 320 nm. Further variable temperature CD/UV experiments down to -70°C on this sample indicate a relative decrease of the intensity of the negative Cotton effect compared with that of the positive Cotton effect.   Thus, at -10°C, two CD bands, one positive and one negative, are apparent, values for both of which are given in Table 1.
50. Fujiki, M.; Toyoda, S.; Yuan, C.-H.; Takigawa, H. *Chirality* **1998**, *10*, 667.
51. Nakanishi, K.; Berova, N. In *Circular Dichroism: Principles and Applications*; Nakanishi, K.; Berova, N.; Woody, R. W., Eds.; VCH: New York, 1994; Chapter 13, p 361.
52. For fluorescence excitation data, see Supplementary Information of refs. 22, 47 and 48.
53. Jha, S. K.; Cheon, K-S.; Green, M. M.; Selinger, J. V. *J. Am. Chem. Soc.* **1999**, *121*, 1665.
54. Green, M. M.; Lifson, S.; Teramoto, A. *Chirality* **1991**, *3*, 285.
55. Bradbury, E. M.; Carpenter, B. G.; Goldman, H. *Biopolymers* **1968**, *6*, 837.
56. Toriumi, H.; Saso, N.; Yasumoto, Y.; Sasaki, S.; Uematsu, I. *Polymer J.* **1979**, *11*, 977.
57. Maeda, K.; Okamoto, Y.; *Macromolecules* **1998**, *31*, 5164.
58. Fujiki, M. *J. Am. Chem. Soc.* **2000**, *122*, 3336.
59. Cheon, K. P.; Selinger, J. V.; Green, M. M.; Angew. Chem. Int. Ed. **2000**, *39*, 1482.
60. Nakashima, H.; Fujiki, M.; Koe, J. R.; Motonaga, M. *J. Am. Chem. Soc.* submitted.
61. Cotts, P.; Miller, R. D.; Sooriyakumaran, R. In *Silicon-Based Polymer Science*; Zeigler, J. M.; Fearon, F. W. G., Eds.; ACS: Washington DC, 1990; Chapter 23, p 397.

62. Estimation of persistence length, q. Using values for Si-Si bond lengths (2.414 Å) and Si-Si-Si bond angles (114.4°) derived from the single crystal X-ray diffraction study of poly(dichlorosilylene) (see Koe, J. R.; Powell, D. R.; Buffy, J. J.; Hayase, S.; West, R. *Angew. Chem. Int. Ed.* **1998**, *37*, 1441) and the viscometric data obtained, an estimate of q may be obtained (see ref. 33; the polymer is treated as Gaussian and expansion factors are not included). For **6**: given the average formula weight per backbone repeat unit of 300.124, the mass per unit length, M_L, is 147.9 Å^{-1}. Considering data for a molecular weight, M, of 1×10^5, the contour length, $L = M/M_L$ is 676 Å. Given also $L = n_k l_k$ (where n_k and l_k are the number of Kuhn segments and the Kuhn segment length, respectively) and the mean-square end-to-end distance $\langle r^2 \rangle = n_k l_k^2$, a value for the Kuhn segment length of the polymer at the chosen molecular weight is given by $l_k = \langle r^2 \rangle / L = 148.65$ Å. Since $l_k = 2q$, the persistence length, q, therefore approximates to 75 Å. Likewise, the persistence length of **1** may be approximated to 45 Å. Considering the approximations made and that the estimate for **1** is shorter than that in ref. 33 (the reported value for **1** is about 100 Å), use of the values derived here for **6** and **1** should be restricted to a qualitative comparison between them. Idealized rigid rod values of R_g for **6** and **1** were obtained by use of calculated L values in the relation for rigid rods: $\langle r^2 \rangle = 12 \langle s^2 \rangle = L^2$, where $\langle s^2 \rangle$ is the mean-square radius of gyration.

63. Fujiki, M. *J. Am. Chem. Soc.* **1996**, *118*, 7424.
64. Fujita, H. *Polymer Solutions*, Elsevier: Amsterdam, 1990.
65. Lovell, P. A. In *Comprehensive Polymer Science*, Allen, G.; Bevington, J. C.; Booth, C.; Price, C., Eds.; Pergamon: Oxford, 1989.

Chapter 7

Structured Polymers with Stimuli-Responsive Chiroptical Behavior: Azobenzene-Modified Helical Constructs

Gary D. Jaycox

DuPont Central Research and Development, E328–205B, Experimental Station, Wilmington, DE 19880–0328

Azobenzene modified polyaramides and several model compounds fitted with atropisomeric 2,2'-binaphthyl linkages exhibited thermo- and photo-responsive chiroptical behavior when evaluated in solution. The *trans*-azobenzene modified polymers afforded CD spectra with intense molar ellipticities in the 300 to 400 nm spectral window. Specific rotation magnitudes at the sodium D-line ranged into the hundreds of degrees and were dependent upon the extent of binaphthyl loading along the polymer chain. The irradiation of the polymer samples to drive the *trans* → *cis* isomerization process resulted in an immediate chiroptical response, with CD band intensities and optical rotations significantly diminished. These effects were fully reversible and were attributed to the presence of one-handed helical conformations in the *trans*-azobenzene modified polymers that were severely disrupted following the *trans* → *cis* isomerization reaction.

Photo- and thermo-regulated conformational changes in azobenzene modified polymers provide an opportunity for the rational design of "smart" materials systems. Particularly noteworthy in this regard have been attempts to utilize local *trans – cis* isomerization reactions to induce well-defined conformational transitions in macromolecules endowed with main chain or pendent side chain azobenzene groups. Light driven α-helix - random coil, α-helix - β-helix and right helix - left helix transitions in azobenzene modified polypeptides have been extensively investigated by Ueno (*1,2*) and Ciardelli (*3-5*). More recently, Zentel and co-workers (*6,7*) have reported on a series of photo-mediated helix inversions in polyisocyanates that have been modified with small quantities of chiral azobenzene side groups.

As depicted schematically in Figure 1, multiple *trans – cis* isomerization reactions triggered within these helical polymers can ultimately give rise to twin photochromic and chiroptical responses that are intimately related. The nature of this inter-relationship can often be fine-tuned by adjusting a number of parameters including polymer backbone structure, the location and extent of azobenzene loading, solvent type and solution pH. Materials of this kind continue to offer considerable insight into a variety of photomodulated biological processes and may be well suited for optical data storage, polymeric sensory devices and other technological applications (*5*).

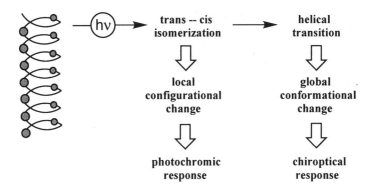

Figure 1. Coupled photochromic and chiroptical responses in azobenzene modified helical polymers.

Efforts within our own laboratory are focused on the development of novel condensation polymers that adopt single-handed helical geometries and also undergo well-defined structural perturbations in response to externally applied stimuli. When viewed generically, our polymers possess conformationally restricted backbones that, minimally, contain two essential components. These include a helical directing group and a stimuli-responsive chromophore or *stimuliphore* (Figure 2). As we (*8-11*) and others (*12*) reported earlier, axially asymmetric R- or S-binaphthyl linkages are particularly effective helical directing groups, generally giving rise to longer range helical order within a polymer backbone when spaced along the main chain at regular intervals. Single

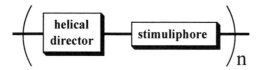

Figure 2. Essential backbone components for stimuli-responsive polymers.

handed helical conformations result when only one optical isomer is utilized to construct the polymer backbone. The incorporation of *trans*-azobenzene, *trans*-stilbene or viologen stimuliphores into properly designed helical backbones provides for new materials in which helical geometries can be readily distorted by the action of externally applied stimuli. Light, heat and red-ox driven changes in chiroptical behavior have all been effectively demonstrated within our systems. In this chapter are presented some of our more recent results in this area, with specific emphasis placed on azobenzene modified helical constructs.

Polymer Synthesis and Physical Properties

The *trans*-azobenzene modified polymers shown in Scheme 1 were prepared in the absence of room light by the solution polycondensation of *trans*-azobenzene-4,4'-dicarbonyl chloride with appropriate diamine monomers. Polymerizations were carried out in DMAC as described earlier (*8,9*).

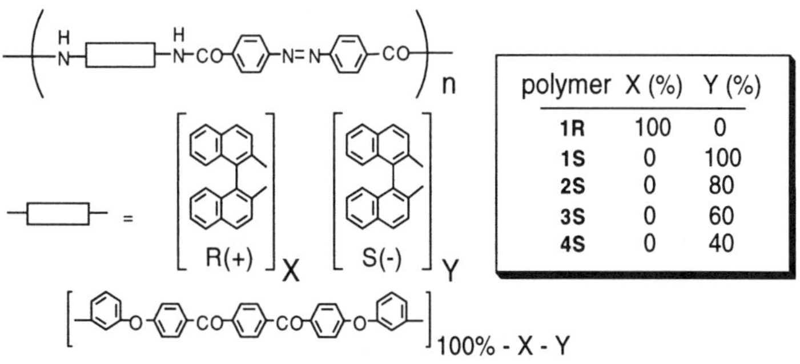

polymer	X (%)	Y (%)
1R	100	0
1S	0	100
2S	0	80
3S	0	60
4S	0	40

Scheme 1. Azobenzene modified polymers.

Polymers **1R – 4S** were soluble in a variety of organic solvents including THF and acetone. Azobenzene contents ranged from about 28 to 35 weight percent while chiral binaphthyl loadings in the polymer backbones varied more broadly, approaching 49 weight percent for polymers **1R** and **1S**. Films cast from DMAC were optically transparent and possessed a good range of physical and mechanical properties. All of these materials exhibited an onset of weight loss under nitrogen or air near 430 °C coincident with the extrusion of molecular nitrogen from the polymer chain (*8,9*). Polymer samples were devoid of crystallinity as evidenced by DSC measurements.

Photophysical Properties

Polymers **1R – 4S** exhibited photo- and thermo-regulated behavior in THF solutions consistent with the presence of multiple azobenzene stimuliphores in their backbones (*8,9*). Photoinduced *trans* → *cis* isomerization reactions were carried out by irradiating polymer solutions with filtered (370 < λ < 400 nm) ultraviolet light. This process afforded photostationary state compositions in which two out of every three azobenzene linkages occupied their higher energy *cis*-configurations. Activation energies for thermal "back" reactions that restored the population of *trans*-azobenzene stimuliphores in the polymer chains fell near 21 to 23 kcal mol^{-1} when evaluated in THF. *Cis*-isomer half-life values at room temperature exceeded one day, enabling the acquisition of chiroptical data over short time periods that were uncomplicated by rapidly shifting *cis – trans* isomer ratios. When desired, thermally driven *cis* → *trans* isomerization reactions in these systems were greatly accelerated by warming the polymer solutions above room temperature. When carried out at 60 °C, nearly all of the more linear *trans*-azobenzene backbone linkages were restored within a matter of several hours. *Cis* → *trans* reorganization along each polymer backbone could also be triggered photochemically by irradiating the polymer solutions with visible light for several minutes. This photo- and thermo-regulated isomerization behavior was completely reversible over numerous irradiation – heating cycles.

Stimuli-Responsive Chiroptical Behavior

The primary motivation for our research program has been focused around the design of high-performance materials systems that exhibit stimuli-responsive chiroptical properties both in dilute solution, and ultimately, in solid state environments. Molecular modeling studies have revealed that *trans*-**1R** and *trans*-**1S** are each capable of occupying chiral, single-handed helical geometries where three *trans*-azobenzene stimuliphores occupy one full turn in the helical screw (*13*). The substitution of more linear aryl ether ketone segments for the atropisomeric binaphthyl moieties in these materials (polymers **2S - 4S**) reduces the number of helical directors present in the main chain, effectively distorting the helical conformations adopted by the polymer coil. The exposure of these polymer solutions to ultraviolet light to promote a series of *trans* → *cis* isomerization reactions along the polymer backbone will further serve to disrupt this helical order. As will be demonstrated below, reversible perturbations in chiral helical geometry can lead to changes in the circular dichroism spectra and optical rotation values measured for this polymer series.

Circular Dichroism Measurements

Multiple circular dichroism (CD) spectra obtained in THF for this polymer series are overlaid in Figure 3 for comparison. As is apparent, *trans*-polymers **1R** and **1S** exhibited mirror image behavior, each showing intense dichroic bands in the 280 to 400 nm spectral window. These bands stem from the *trans*-azobenzene π,π^* transition and are clearly split into two distinct regions having both positive and negative amplitudes. The zero points located near 338 nm in Figure 3 correspond the wavelength where the *trans*-azobenzene stimuliphore possesses a maximum in its optical absorbance spectrum. CD line shapes for the other *trans*-polymer variants were qualitatively similar. Bisignated dichroic spectra like those shown in Figure 3 have been observed for other azobenzene modified helical polymers that occupy single-handed conformations in solution (*7,12,14*). This behavior has been attributed to exciton coupling that arises in response to superstructures that position two (or more) *trans*-azobenzene groups in chiral space such that dipole-dipole interactions between neighboring stimuliphores become possible.

Exposing these polymer solutions to filtered ($370 < \lambda < 400$ nm) ultra-violet light afforded a dramatic change in their CD spectra. Specifically, CD band intensities were significantly reduced within the 280 to 400 nm spectral window. A CD spectrum for polymer **1R** having a light-induced *cis*- to *trans*-azobenzene backbone content near 65 percent - 35 percent is provided in Figure 4 (plot T_o). Here, molar ellipticity values for the twin extrema centered at 318 and 359 nm were reduced by a factor of three following the illumination step. As expected, this light driven chiroptical response was completely reversible. Time-resolved CD spectra obtained at various intervals following the illumination step are also given in Figure 4 for comparison. Notably, a CD plot acquired after 214 hours of "dark" exposure was virtually identical to a spectrum obtained for the pre-irradiated polymer (plot *trans*-**1R**).

This reversible chiroptical response arises from a combination of local and longer range effects that are likely to be operative within these systems. The azobenzene stimuliphores in *trans*-**1R** are flanked on either side by neighboring chiral binaphthyl segments and thus reside locally in an asymmetric environment. This environment may be further perturbed by the presence of longer range helical conformations that are also chiral. The light mediated *trans* → *cis* isomerization reaction triggered within the polymer's main chain will significantly diminish the population of *trans*-azobenzene linkages that give rise to the π,π^* band centered at 338 nm. This local photochromic response will reduce the magnitude of the bisignated Cotton effects appearing between 280 and 400 nm in Figure 4. At the same time, this isomerization process will also

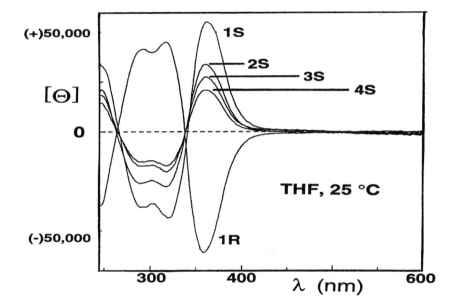

Figure 3. CD spectra for trans-polymer series.

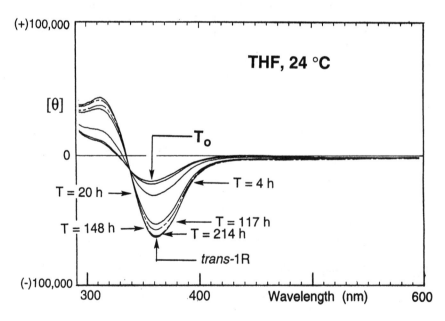

*Figure 4. Time-resolved CD spectra for polymer **1R** following the illumination step.*

disrupt the global helical conformations that have been predicted for the polymer backbone, potentially modifying further the CD response for this polymer chain. A series of thermally mediated *cis → trans* relaxation steps triggered within the polymer will simultaneously restore both the number of *trans*-azobenzene groups and the helical geometries that they are likely to induce, effectively driving the chiroptical response depicted in Figure 4. That the CD spectrum for polymer **1R** obtained after 214 hours in the dark was nearly identical to a plot for the pre-irradiated polymer sample is fully consistent with a *cis*-azobenzene half-life near 50 hours determined independently by optical absorbance spectroscopy under these experimental conditions.

Optical Rotation Measurements

Additional insight into the nature of the stimuli-responsive chiroptical properties displayed by this polymer series was gained by carrying out optical rotation measurements in THF. When recorded at the sodium D-line, these measurements were well removed from the strong π,π^* and much weaker n,π^* azobenzene transitions localized within the polymer backbones. Specific rotation values measured in THF before and after ultraviolet light exposure are displayed graphically in Figure 5 for comparison. Data for polymer **4S** were not included due to this variant's limited solubility in the THF solvent medium.

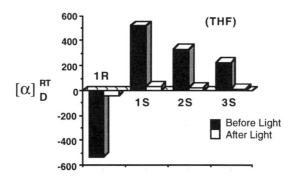

Figure 5. D-line specific rotations before and after ultraviolet light exposure.

The optical rotation results displayed in Figure 5 are interesting in several respects. Firstly, in the pre-irradiated conformational states, all of the polymers modified with the binaphthyl helical director exhibited large specific rotations in THF, with rotation magnitudes trending upward with increasing binaphthyl monomer contents. Secondly, polymers **1R** and **1S** were characterized by mirror image behavior consistent with expectations derived from molecular modeling studies which predicted that the two polymers should adopt opposite helical screw directions. Finally, the illumination of the four polymer samples with filtered ultraviolet light produced an immediate drop in their optical rotatory powers. In all cases, rotation values were reduced by more than a full order of magnitude. These results parallel those obtained from CD measurements described in the section above. The light-induced isomerization of the *trans*-azobenzene stimuliphores present in these materials results in a distortion of chiral helical conformations, providing for dramatic changes in chiroptical behavior. That macromolecules with single handed helical geometries can exhibit large optical rotations in solution and in the solid state is now widely recognized (*15,16*). D-line specific rotations reaching into the hundreds and even thousands of degrees have been reported for a variety of synthetic polymers, including polychloral (*17,18*), some poly(tritylmethacrylate)s (*19,20*), and several poly(isocyanide)s (*21,22*) that possess chirality solely at the helical or macromolecular level.

The specific rotation values for polymer **1S** in DMAC were also tracked as a function of time following the irradiation step. Changes in this polymer's optical rotatory power were evaluated at three different isotherms for a sample that was stored in the dark to promote the thermally mediated *cis* → *trans* "back" reaction. As demonstrated by the linear plots provided in Figure 6, the recovery

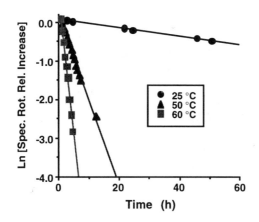

Figure 6. First-order plots for recovery of specific rotations in 1S.

of the polymer's rotation magnitudes was clearly a first-order process that was characterized by a strong temperature dependence. Rate constants calculated for the restoration of optical activity in **1S** varied from 1.86×10^{-4} min^{-1} at 25 °C to a velocity nearly 65 times faster at the 60 °C isotherm. Based on the polarimetric data shown in Figure 6, an activation energy for the restoration of optical activity in polymer **1S** was calculated to fall near 23 kcal mol[-1]. This value is in excellent agreement with an activation energy determined for *cis → trans* return in **1S** by proton NMR and optical absorbance spectroscopies *(9,23)*. Notably, the isomerization process predicted to restore helical backbone structure and the optical rotatory power exhibited by the polymer sample are thus well correlated. This finding is entirely consistent with the explanation provided above for the origins of optical activity in these stimuli-responsive polymers.

Model Compound Studies

Three individual model compounds representing fragments of the polymer backbone in **1R** were also evaluated for their chiroptical behavior as part of this study. Derivatives **5R**, **6R** and **7RR** were constructed from the R-binapthyl-2,2'-diamine building block as described earlier *(13)* and are shown in Scheme 2 below.

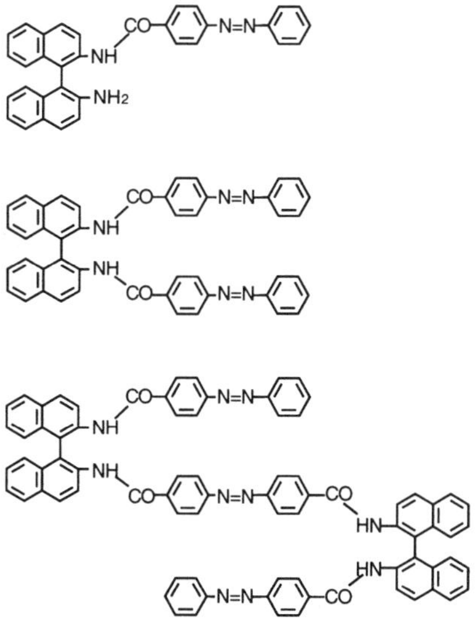

Scheme 2. Model compounds prepared from the R-binaphthyl helical director.

Derivatives **6R** and **7RR** exhibited CD spectra in THF that were nearly identical to that recorded for polymer **1R**. Each *trans*-derivative was characterized by a negative first Cotton effect with a positive second Cotton effect appearing at higher energies. Zero points in each bisignated spectrum were centered near 328 nm. As expected, model *trans*-**5R** fitted with only one *trans*-azobenzene "arm" provided for radically different CD behavior. In this case, a bisignated CD spectrum was not observed due to a lack of exciton coupling between multiple azobenzene stimuliphores within the same molecule. Predictably, the exposure of each solution to filtered ultraviolet light brought about an immediate reduction in CD band intensities. As was the case for their polymeric analogue **1R**, these effects were fully reversible over multiple irradiation – heating cycles.

Specific rotation measurements gathered for the three model compounds offered several surprises. Rotation data acquired in THF are provided graphically in Figure 7. Results for polymer **1R** have also been included for comparison.

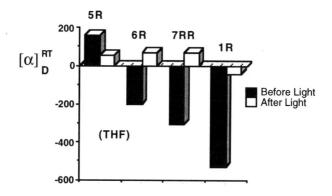

Figure 7. D-line specific rotations before and after ultraviolet light exposure for the three model compounds and their polymeric analogue.

As is readily apparent from the figure provided above, *trans*-models **6R** and **7RR** were characterized by strong levorotatory behavior, with rotation magnitudes clearly scaling with increasing "chain length". Derivative *trans*-**7RR**, representing nearly one full turn in the polymer's helical coil based on preliminary modeling studies, afforded a D-line specific rotation near (-) 308 deg dm^{-1} g^{-1} cm^3. This value was reduced by a factor of 1.7 when compared to that measured for the larger polymer coil. This result suggests that optical activity builds rather quickly in these structures and that only a limited number of helical turns are required before large rotation magnitudes are achieved.

Interestingly, model *trans*-**5R** exhibited a positive specific rotation of 158 deg dm^{-1} g^{-1} cm^{3} that was nearly identical to the rotation value determined for the R-(+)-1,1'-binaphthyl-2,2'-diamine building block used to construct these systems.

Exposing the three model compounds to filtered ultraviolet light produced an immediate drop in their optical rotatory powers, again consistent with the behavior of their larger polymeric analogues. However, in stark contrast to polymer **1R**, the relative signs of rotation for **6R** and **7RR** were inverted following the illumination step (Figure 7). The twin model compounds thus join a unique class of materials that can be reversibly switched from one state to another in response to an applied stimulus (*24*). When compared to polymer **1R**, models **6R** and **7RR** are capable of reaching much "deeper" photostationary states, with approximately 85 to 90 percent of the azobenzene groups in these structures ultimately occupying the higher energy *cis*-configuration. By comparison, photostationary states in polymer **1R** were considerably less enriched in the *cis*-isomer as evidenced by tandem NMR and optical absorbance spectroscopies (*9,23*). This disparity is most likely due to steric constraints imposed by the larger polymer chain that effectively limit the extent of the *trans* → *cis* isomerization reaction.

This light-triggered inversion of chiroptical behavior was investigated more extensively for the twin "armed" model compound **6R**. A solution of this *trans*-derivative in THF was irradiated first with filtered ultraviolet light to effect the *trans* → *cis* isomerization reaction and then with visible light to photochemically induce *cis* → *trans* return of the pendent azobenzene groups. As shown in Figure 8, cycling between the two applied light frequencies resulted in the reversible chiroptical switching of the model compound. This stimuli-responsive behavior was completely reproducible over multiple irradiation cycles with no apparent fatigue or hysteresis effects. Similar results were obtained for model **7RR**.

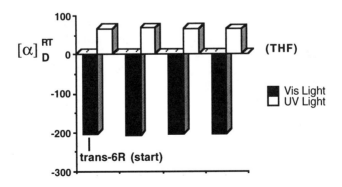

*Figure 8. Model **6R** as a light-driven chiroptical switch.*

Summary and Future Outlook

The covalent linkage of azobenzene stimuliphores to axially asymmetric binaphthyl helical directors has provided for a new series of high and low molecular weight species that exhibit stimuli-responsive chiroptical behavior in dilute solution. These robust derivatives are readily accessible through relatively simple synthetic procedures and are representative of a larger family of structured helical constructs that are under development in our laboratory.

Present efforts are focused on the extension of this work to solid state systems that may be better suited for a variety of novel sensory and switching applications. The incorporation of some of these derivatives into low Tg polymer matrices is one approach under consideration. Structured polymers like *trans*-1R and *trans*-1S can also serve as helical scaffolds or templates, allowing for the secure placement of a variety of functional groups into well defined chiral space.

The work described in this chapter has focused exclusively on the azo-benzene stimuliphore derived from *trans*-azobenzene-4,4'-dicarbonyl chloride. Of course, other stimuli-responsive moieties are also available, including "push-pull" azobenzenes, stilbenes, spirobenzopyrans and viologen chromophores to name a few. Their inclusion into helical structures of this kind would clearly provide for new optically active materials that are sensitive to a variety of different stimuli. Helical constructs fashioned from two different stimuliphoric components are also worthy targets for further exploration. Such systems now under development in our laboratory could potentially exhibit multi-dimensional sensory behaviors. Details concerning some of these efforts will be described in a number of forthcoming reports.

Acknowledgments

The author would like to acknowledge Gerry J. Everlof, Steve R. Lustig and Laurie A. Howe for their technical assistance. Julie A. Ferguson is also thanked for her help with the preparation of this manuscript. This work is part of a larger program on Stimuli-Responsive Polymers and is DuPont contribution number 8057.

References

1. Ueno, A.; Anzai, J.; Osa, T.; Kadoma, Y. *Bull. Chem. Soc. Japan* **1979**, *52*, 549.
2. Ueno, A.; Takahashi, K.; Anzai, J.; Osa, T. *J. Am. Chem. Soc.* **1981**, *103*, 6410.

3. Pieroni, O.; Houben, J.L.; Fissi, A.; Costantino, P.; Ciardelli, F. *J. Am. Chem. Soc.* **1980**, *102*, 5913.
4. Fissi, A; Pieroni, O.; Ciardelli, F. *Biopolymers* **1987**, *26*, 1993.
5. Ciardelli, F.; Pieroni, O.; Fissi, A.; Carlini, C.; Altomare, A. *Brit. Polymer J.* **1989**, *21*, 97.
6. Maxein, G.; Zentel, R. *Macromolecules* **1995**, *28*, 8438.
7. Muller, M.; Zentel, R. *Macromolecules* **1996**, *29*, 1609.
8. Jaycox, G.D. *Polym. Prepr., Am. Chem. Soc., Div. Polym. Chem.* **1998**, *39(2),* 472.
9. Howe, L.A.; Jaycox, G.D. *J. Polym. Sci., Polym. Chem. Ed.* **1998**, *36*, 2827.
10. Jaycox, G.D.; Everlof, G.J. *Polym. Prepr., Am. Chem. Soc., Div. Polym. Chem.* **1999**, *40(1),* 536.
11. Everlof, G.J.; Jaycox, G.D. *Polymer* **2000**, *41*, 6527.
12. Kondo, F.; Takahashi, D.; Kimura, H.; Takeishi, M. *Polym. J.* **1998**, *30*, 161.
13. Lustig, S.R.; Everlof, G.J.; Jaycox, G.D. *Macromolecules*, submitted.
14. Kondo, F.; Hidaka, M.; Kakimi, S.; Kimura, H.; Takeishi, M. *Polym. Prepr., Am. Chem. Soc., Div. Polym. Chem* **1997**, *38(2)*, 209.
15. Vogl, O.; Jaycox, G.D. *Polymer*, **1987**, *28*, 2179.
16. Vogl, O.; Jaycox, G.D.; Kratky, C.; Simonsick, Jr., W.J.; Hatada, K. *Accts. Chem. Res.*, **1992**, *25*, 408.
17. Jaycox, G.D.; Vogl, O. *Polym. J.*, **1991**, *23*, 1213.
18. Jaycox, G.D.; Vogl, O. *Makromol. Chem., Rapid Commun.*, **1990**, *11*, 61.
19. Okamoto, Y.; Suzuki, K.; Ohta, K.; Hatada, K.; Yuki, H. *J. Am. Chem. Soc.*, **1979**, *101*, 4763.
20. Okamoto, Y.; Yashima, E.; Nakano, T.; Hatada, K. *Chem. Lett.*, **1987**, 759.
21. Nolte, R.J.M.; van Beijnen, A.J.M.; Drenth, W. *J. Am. Chem. Soc.*, **1974**, *96*, 5932.
22. Drenth, W.; Nolte, R.J.M. *Accts. Chem. Res.*, **1979**, *12*, 30.
23. Jaycox, G.D.; Howe, L.A.; Beattie, M.S. *Polym. Prepr., Am. Chem. Soc., Div. Polym. Chem.* **1998**, *39(2),* 332.
24. Feringa, B.L.; Huck, N.P.M.; Schoevaars, A.M. *Adv. Mater.*, **1996**, *8*, 681.

Synthetic Peptides

Chapter 8

Collagen Mimics: Synthesis and Properties of Disulfide-Bridged Trimeric Collagen Peptides

Luis Moroder, Stella Fiori, Rainer Friedrich, J. Constanze D. Müller, and Johannes Ottl

Max-Planck-Institute of Biochemistry, 82152 Martinsried, Germany

Heterotrimeric collagenous peptides consisting of (Gly-Pro-Hyp)$_n$ repeats (n = 4 or 5) were assembled in a defined register of the three α chains by the use of a simple cystine knot connecting via a disulfide bridge α1 with α2 and α2 with α1′, respectively. Conformational analysis confirmed a stable triple helix for the trimer with n = 5, thus supporting a role of the cystine knot in induction and stabilization of the collagen-type triple-helical fold. This strategy was applied for the synthesis of mimics of the collagenase cleavage site of collagen type I. The selective digestion of the heterotrimers by collagenases, besides revealing two distinct mechanisms of proteolysis by collagenases and gelatinases, confirmed that the α1α2α1′ alignment of the chains most probably represents the natural register of the α chains at least of collagen type I.

Collagen is the most abundant tissue protein that serves as structural material, but also as anchorage for all stationary cells and as a track for cell migration. These proteins are structurally rod-shaped macromolecules consisting of three identical or of two or three different α chains of primarily repeating Gly-Xaa-Yaa triplets. The high content of proline and hydroxyproline in positions Xaa and Yaa, respectively, and the presence of a glycine at every third sequence position induces each α-chain to adopt a left-handed poly-Pro-II helix and the three chains to intertwine with a one-residue shift into a right-handed triple-helical supercoil. This unique collagen structure is stabilized by an extensive hydrogen bond and hydration network (for recent reviews see ref. 1-3). The α-chains are synthesized *in vivo* as precursor molecules that contain globular domains at both the N- and C-termini. The assembly of the α-subunits in the correct register is induced by a selective recognition of the C-propeptides (4), whereupon nucleation of the triple helix takes place at this C-terminus and the supercoiling is supposed to propagate in the C→N direction in a zipper-like mode with a rate that is limited mainly by the slow *cis→trans* isomerization of Xaa-Pro imide bonds (5). Upon maturation, the homo- or heterotrimeric macromolecules are stabilized in some collagens by complex cystine networks, although interchain-disulfide bridging is not a prerequisite for correct assembly (6, 7).

Due to the characteristic physical properties of collagens, in particular of the fibrillar type I, II and III collagens, intensive research has been performed in the past decades with synthetic collagen peptides (8) and bioexpressed collagen mutants (9, 10) to understand the optimal amino acid composition and to investigate into details the structural parameters responsible for the triple-helical fold.

Synthesis of Collagen Peptides

Early synthetic work on collagen peptides provided clear evidence that optimal sequence repeats in sequential oligopeptides as mimics of collagen are $(Gly-Pro-Hyp)_n$ and that a peptide length of $n = 10$ is required for a stable triple helix in aqueous solution and at room temperature. Moreover, replacement of Hyp with Pro and in more pronounced manner of Hyp and/or Pro with other amino acid residues significantly decreases the stability of the triple helix (2, 8, 11). More recently it was shown that even replacements of single residues in $(Gly-Pro-Hyp)_{10}$ clearly affect the collagen structure, thus yielding a better understanding of genetic diseases related to single-point mutations in natural collagens (2). Folding and unfolding studies with both synthetic and natural collagen fragments confirmed the entropic penalty of the association of single-stranded collagenous peptides into triple-helical homotrimers. In fact, refolding of larger natural sequences may require days or even fail (12). Conversely, the refolding process occurs at high rates if natural collagen fragments are used which are crosslinked by cystine knots (12, 13). To overcome the unfavorable entropy of self-association of single-stranded collagenous peptides and to possibly facilitate nucleation of the triple helix, their N- and/or C-terminal crosslinking with homotrifunctional templates into homotrimers has been extensively applied, whereby spacers differing in length and flexibility were adopted to allow a staggered alignment of the three chains in a triple-helical fold (Figure 1). More recently, the conformationally constrained Kemp's triacid, i.e. *cis,cis*-1,3,5-trimethylcyclohexane-1,3,5-tricarboxylic acid (Figure 1, compound VI), with its carboxyl groups aligned in parallel orientation was found to represent a very efficient template in terms of triple helix nucleation and stabilization (14-16). In fact, with a N-terminally crosslinked (Gly-Pro-Hyp)$_3$ construct an incipient triple-helical fold was observed, and with an (Gly-Pro-Hyp)$_5$ crosslinked homotrimer a fully stable collagen-like structure was obtained.

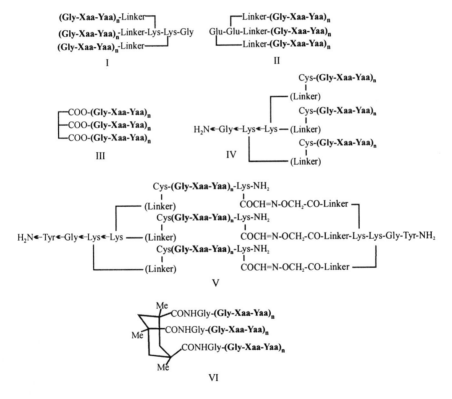

Figure 1. Templates used for the assembly of homotrimeric collagenous molecules: I (17-20), II (21), III (17, 18), IV (20), V (22) and VI (14-16); the arrows in compound IV and V indicate the N→C peptide bond direction.

Design of Heterotrimeric Collagen Peptides

In nature, a turnover of the extracellular matrix is required in all remodeling processes of the connective tissues during growth and development, but it is also associated with serious pathophysiological processes such as cancer cell invasion and thus, tumor metastasis (23). The fibrillar type I, II and III interstitial collagens are digested by vertebrate collagenases, i.e. by MMP-1, MMP-8 and MMP-13, at appreciable rates under physiological conditions, although this proteolytic process still remains the slowest enzymatic substrate degradation known so far. The collagenases cleave the rod-like molecules in a highly specific manner, by a single cut across all three α chains of the collagens at distinct sensitive loci generating characteristic ¾ and ¼ fragments (24).

If mimics of functional collagen epitopes such as the collagenase-cleavage site of collagen type I or the integrin-recognition and -binding site of collagen type IV are the goal, heterotrimeric constructs have to be designed that are synthetically accessible. Crosslinking of different peptide chains on the templates shown in Figure 1, to produce heterotrimers would require a highly selective and sophisticated chemistry, and thus efficient syntheses are difficult to achieve (25).

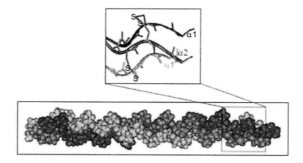

Figure 2. CPK model of the (Gly-Pro-Hyp)$_n$ triple helix with the built-in C-terminal cystine-knot (insert).

We have attempted an approach which is based on a mimic of the natural cystine knots of collagens. Thereby the main problem of how to assemble heterotrimeric constructs in a regioselective manner with induction of a one-residue shift of the three peptide chains in the correct register had to be solved. For this purpose the structural model of the (Gly-Pro-Hyp)$_{10}$ triple helix was built up using the coordinates derived from X-ray crystallographic analysis of this homotrimeric complex (26, 27). The C-terminal Gly-Pro-Hyp repeat was replaced with Gly-Cys-Gly in the α1, with Gly-Cys-Cys in the α2 and with Gly-

Pro-Cys in the α1′ chain and the triple helix was crosslinked with a simple cystine knot consisting of two interchain disulfide bridges connecting the α1 to α2, and α2 to α1′ peptide, respectively. After energy minimization of this model structure, a surprisingly good fitting of the two disulfide groups into the triple-helical fold was observed, as shown by the insert of Figure 2, and no steric clashes were detected.

Synthesis of Triple-Stranded Cystine-Peptides

Although the chemistry available for regioselective disulfide bridging of cysteine peptides allows for unambiguous synthesis of heterodimers, direct extension of this existing methodology to the regioselective assembly of three different cysteine peptides (α1, α2 and α1′) into a heterotrimer proved to be exceedingly difficult (28). To build-up the C-terminal artificial cystine knot between three collagenous peptides, a selective chemistry had to be elaborated that strictly prevents undesired thiol/disulfide exchange reactions. For such purpose the synthetic route outlined in Figure 3 was developed (29).

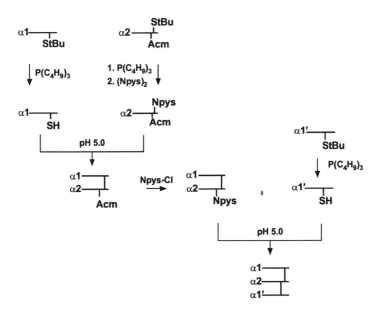

Figure 3. Regioselective assembly of the α1, α2 and α1′ cysteine peptides into heterotrimers with an α1α2α1′ register.

The thiol-activation and disulfide-formation steps are performed at pH 4.5-5.5 where thiol/disulfide exchange reactions at already formed disulfide bridges and homodimerization of cysteine-unprotected intermediates are largely prevented if the due precautions are taken to avoid contact with air-oxygen. Heterotrimers were obtained by this procedure which differ in size (up to 10 kDa) and sequence composition and where the single peptide chains are aligned in a defined order to form with an one-residue shift the right-handed triple helix as the characteristic collagen structure (30-32).

Heterotrimeric Collagen Peptides of the (Pro-Hyp-Gly)$_n$ Type

To experimentally analyze the contribution of the cystine knot to nucleation and stabilization of the triple helix, the two heterotrimers **1** and **2** with 4 and 5 (Pro-Hyp-Gly) repeats (Figure 4) were synthesized following the synthetic strategy outlined in Figure 3 and using Fmoc-Pro-Hyp-Gly-OH as synthons. The ratios of the dichroic intensities of the positive over the negative maximum (Rpn values as indices of the content of triple-helical conformation; see ref. 16) as derived from the CD spectra of both constructs in a 4:1 water/ethylene glycol mixture (the glycol is known to stabilize the triple helix; see ref. 33) at 4 °C show values > 0.1 confirming a collagen-type fold; this observation agrees with similar constructs obtained with the rigid Kemp's triacid template (15, 34). However, the time requirement for conformational stabilization at 4 °C differs for the two trimers as monitored by NMR spectroscopy, where trimer **1** takes days to reach the equilibrium, whilst the trimer **2** upon dissolution needs only 2 h storage at this low temperature. NMR confomational analysis of the trimer **2** at 4 °C revealed a single set of signals for the component amino acid residues indicating that the whole molecule is folded into the triple helix. The results strongly support the working assumption that the cystine knot is capable of efficiently nucleate and stabilize the triple helix in the model peptides, despite its lack of rigidity.

$$Ac\text{-}(POG)_n\,G\text{-}OH$$
$$Ac\text{-}(POG)_n\overset{|}{C}CG\text{-}OH$$
$$H\text{-}(POG)\ P\overset{|}{C}G\text{-}OH$$

*Figure 4. Sequence composition of two trimeric model collagen peptides with a C-terminal cystine knot: trimer **1** (n = 4) and trimer **2** (n = 5).*

Heterotrimeric Collagen Peptides Related to the Collagenase Cleavage Site of Collagen Type I

According to the existing knowledge on the sequence-dependent stability of triple helices (5), the low content of imino acids that characterizes the loci of collagen cleavage by collagenases, most probably serves to untighten the triple-helical fold in this portion of the molecule and thus to generate a conformational epitope that is specifically recognized and cleaved by the interstitial collagenases. To investigate the conformational properties of this epitope,

heterotrimer A
H-(GPO)$_3$GPQG|IAGQRGVVGCGG-OH
H-(GPO)$_3$GPQG|LLGAOGILGCCGG-OH
H-(GPO)$_3$GPQG|IAGQRGVVGLCGG-OH

heterotrimer B
H-(GPO)$_5$GPQG|IAGQRGVVGCGG-OH
H-(GPO)$_5$GPQG|LLGAOGILGCCGG-OH
H-(GPO)$_5$GPQG|IAGQRGVVGLCGG-OH

heterotrimer C
Ac-PO(GPO)$_5$GPQG|IAGQRGVVGPOGCG-OH
Ac-O(GPO)$_5$GPQG|LLGAOGILGPOGCCG-OH
H-(GPO)$_5$GPQG|IAGQRGVVGPOGPCG-OH

heterotrimer D
Dns-PO(GPO)$_5$GPQG|IAGNRGVVGCG-OH
H-O(GPO)$_5$GPQG|LLGAOGILGCCG-OH
H-W-(GPO) GPQG|IAGQRGVVGLCG-OH

Figure 5. Sequence composition of the synthetic heterotrimers as mimics of the collagenase cleavage site of collagen type I. Enzymatic cleavage in collagen occurs at the Gly^{775}-Ile^{776} bond of the α1 and α1' chains and at the Gly^{775}-Leu^{776} bond of the α2 chain as indicated by the line.

synthetic heterotrimeric collagen peptides were synthesized which contain the collagenase cleavage site-sequences 772-784 (P$_4$-P$_9$·) and 772-785 (P$_4$-P$_{10}$·) of the two α1 chains and the cleavage site-sequence 772-784 (P$_4$-P$_9$·) of the α2 chain of collagen type I. The natural sequences were crosslinked at the C-terminus with the cystine knot as discussed for the trimeric model collagen peptides and extended N-terminally with (Gly-Pro-Hyp)$_n$ repeats to stabilize the hypothetically weak triple helix of the natural digestion epitope (30, 31). The display of the amino acid side chains in the heterotrimeric epitope is crucially affected by the stagger of the three α chains. For collagen type I an α1α2α1' register has been proposed on the basis of modeling experiments (35), but unambiguous experimental proofs for such alignment of the chains were missing.

Nonetheless, this type of chain assembly was selected for the synthetic heterotrimers shown in Figure 5.

Triple Helix Stability of the Heterotrimeric Collagen Peptides

The heterotrimers A, B, C and D of Figure 5 exhibit at 4 °C CD spectra that are characteristic of a triple-helical conformation, and the Rpn values are supportive of high contents of the supercoiled fold (Table I). Conversely, the Rpn values of the single-chain peptides are compatible only with their partial folding into poly-Pro-II helices (31). This observation agrees with the supposedly weak triple-helix propensity of these sequence portions of collagen type I, despite the presence of a glycine residue at every third position and of a length of the single chains (30 to 33 residues) that generally suffices for full stabilization of a triple-helical fold of non-crosslinked collagenous peptides as well assessed with the sequential polymers $(Gly-Pro-Hyp)_{10}$ and $(Gly-Pro-Pro)_{10}$ (8).

Table I. CD parameters of the synthetic heterotrimeric collagen peptides*.

molecules	max (nm, Θ_R)	min (nm, Θ_R)	Rpn	T_m (°C)
heterotrimer A	222 (2100)	199 (-18200)	0.115	9
heterotrimer B	222 (4160)	196 (-34000)	0.122	33
heterotrimer C	223 (4350)	197 (-33700)	0.129	41
heterotrimer D	224 (2920)	198 (-33600)	0.087	30
collagen type I	222 (4500)	196 (-34600)	0.130	38

*) The CD spectra were recorded in the enzymatic assay buffer (50 mM Tris·HCl, pH 7.4, 10 mM $CaCl_2$, 50 mM NaCl) at 4 °C.

Thermal unfolding and refolding of the heterotrimers as monitored by CD at 222 nm, clearly revealed that this process is fully reversible and highly cooperative with melting temperatures (T_m) affected in decisive manner by the size of the $(Gly-Pro-Hyp)_n$ extension. N-Terminal acetylation of two chains to suppress electrostatic repulsions enhances the stability of the heterotrimers to some extent. Moreover, by incorporation of an additional (Gly-Pro-Hyp) repeat adjacent to the cystine knot and by filling with residues the sequential gaps deriving from the one-residue shift of the three α chains in the triple helix, a significant stabilization of the supercoil is induced which in trimer C reaches a T_m value similar to that of natural collagen (Table I).

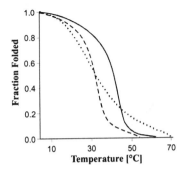

Figure 6. Thermal unfolding of the trimers B (- - - - -), C (———) and D·(•••••) in 50 mM Tris·HCl, pH 7.4, 10 mM CaCl₂, 50 mM NaCl.

Although the heterotrimers consist of two distinct sequence portions, i.e. the N-terminal (Gly-Pro-Hyp) repeats with optimal and the C-terminal (Gly-Xaa-Yaa) repeats with weak triple-helix propensity, a two-phase transition is not observed as exemplarily shown in Figure 6 for the more stable trimers B and C. Two-phase thermal transitions in collagens with triple-helical domains of different sequence composition are rarely observed and apparently only in those special cases were the stability of the single domains differs significantly (36). The molar ellipticity values expressed per residue ($[\Theta]_R$) of the trimers B, C and D are similar to those determined for model collagen peptides where NMR and X-ray crystallographic analysis confirmed a triple-helical fold involving the entire chain length. This would strongly support the onset of a triple-helix spanning the heterotrimers from the C- to the N-terminus and thus, a more or less cooperative unfolding of the whole molecule including the natural collagenase-cleavage epitope of collagen type I.

Digestion of the Heterotrimeric Collagen Peptides by Collagenases and Gelatinases

The heterotrimer A with its T_m value of 9 °C is fully unfolded at room temperature and thus gelatin-like in the enzyme buffer, whilst the trimers B (T_m = 33 °C) and C (T_m = 41 °C) retain the collagen-like structure under these conditions. Correspondingly, these two types of trimers were assayed as mimics of gelatin and collagen for their catabolism by gelatinases and collagenases, respectively. In contrast to the difficult handling of natural collagen substrates (37), the high water-solubility and relatively low size of the synthetic trimers allowed to follow the enzyme kinetics in efficient manner by HPLC (Figure 7).

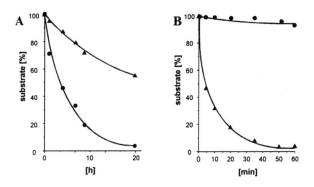

Figure 7. Digestion of the heterotrimers A (▲) and C (●) by A) the collagenases MMP-1 and MMP-8 and B) the gelatinase MMP-2 as monitored by HPLC.

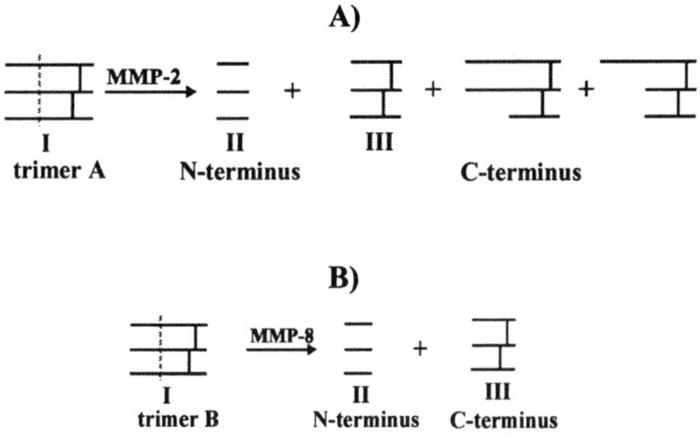

A)

I
trimer A   II   III
          N-terminus   C-terminus

B)

I
trimer B   II   III
          N-terminus   C-terminus

*Figure 8. Product distribution of the proteolytic degradation of collagen- and gelatin-like substrates by (**A**) the gelatinase MMP-2 and (**B**) the collagenases MMP-1 and MMP-8.*

The trimer A as a gelatin-like substrate is cleaved by the collagenases MMP-1 and MMP-8 at very slow rates, whilst the collagen-like substrate C is digested at comparatively high rates. An opposite picture was obtained with gelatinase A (MMP-2) which cleaves at high rates the gelatin-like trimer A, but at markedly lower rates the triple-helical substrates B and C (30, 38). Taking into account that the display of the amino acid side chains differs strongly upon the register of the three α chains, the specific recognition of the trimers B and C, assembled as $\alpha 1 \alpha 2 \alpha 1'$ trimers, by the collagenases strongly suggests that this chain alignment corresponds to the natural register of collagen type I. The results also confirm that the interstitial collagenases are the responsible enzymes for the catabolism of native collagen and that gelatinases are mainly involved in the processing of denatured collagen, i.e. of gelatin (39).

LC-MS analysis of the product distribution in the time course of enzymatic degradation of the synthetic substrates led to the additional important discovery that the two enzymes not only operate in strong conformation-dependency of the substrates, but also by two distinct proteolytic mechanisms (38). Digestion of the trimeric substrates by gelatinase occurs in individual steps, with release of partially digested trimers into the medium, independently of whether gelatin- or collagen-like substrate is cleaved (Figure 8A). Conversely, both types of substrates are trapped by the collagenases until scission through all three α chains is completed (Figure 8B). This is a time-requiring process involving a potential unfolding of at least the triple-helical substrate and the successive presentation of the three α-chains to the active site.

Fluorogenic Triple-Helical Substrates of Collagenases

In view of the results obtained with the heterotrimeric collagenous peptides as mimics of the natural substrates collagen and gelatin, the design of a fluorogenic substrate was attempted by exploiting a conformation-dependent fluorescence resonance energy transfer (FRET) for quenching the fluorescence in an intact triple-helical heterotrimer. Upon enzymatic digestion the fluorescence intensity should increase and thus allow spectroscopic monitoring of substrate proteolysis. The distances between the Cα atoms of the amino acids located N-terminally on a vertical cut through the collagen triple-helix range between 4.7 to 4.9 Å. Thus incorporation of the pair of chromophores tryptophan/dansyl (critical energy transfer distance $R_0 = 21.3$ Å for FRET) at the N-termini of two α chains of a heterotrimeric construct was expected to fulfil all requirements for optimal energy transfer if the triple helix propagates up to the N-terminus to constrain the two chromophores in close spatial proximity. For this purpose the heterotrimer D of Figure 5 was synthesized where the gaps of the trimer B resulting from the one-residue shift of the α1α2α1′ alignment were occupied in sequence mode with amino acid residues and the N-termini of the α1 and α1′ were dansylated and acylated with the tryptophan residue, respectively (32). Incorporation of the two hydrophobic residues at the N-terminus of the trimer was expected to additionally stabilize the triple helix via hydrophobic interactions between the indole moiety and the dansyl group in view of previous results with amphiphilic collagenous peptides (40). Apparently, the opposite effect was induced, i.e. a destabilization of the triple helix as well assessed by the lower Rpn and T_m values of the heterotrimer D (Table I). Moreover, a

114

significantly less cooperative thermal transition is apparently induced by the N-terminal destabilization (Figure 6). The enhanced mobility of the N-termini is also confirmed by the average distances between the two chromophores as determined by FRET. By modeling a rigid triple helix with the N-terminal chromophores and allowing for free rotation of the indole and dansyl group the distance between the pseudoatoms of the aromatics ranges from 8 Å to a maximum of 12 Å. The average distance value derived from the experimental fluorescence intensities of the tryptophan residue and referred to Ac-Trp-NH$_2$ as an non-quenched probe is 15-16 Å at 4 °C. The identical values were obtained at 23 °C, i. e. at 7 °C below the T$_m$ value of this trimer. The N-terminal chromophores are, therefore, sufficiently restricted in their conformational space to allow fluorescence polarization as well as fluorescence quenching to be exploited for monitoring enzymatic proteolysis and thus disrupture of the triple helix (Figure 9). Thereby for fluorescence quenching measurements, cooling of aliquots of the digestion mixture to 4 °C was found to enhance the difference in fluorescence intensity between intact substrate and digestion mixture.

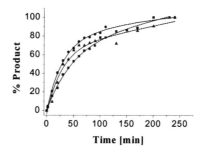

Figure 9. Enzymatic digestion of the fluorogenic triple-helical trimer D by MMP-13 at 23 °C at a substrate/enzyme ratio of 100:1 as monitored comparatively by HPLC (●), fluorescence intensity (▲) and fluorescence polarization (■).

Conclusions

The results obtained with the disulfide-bridged heterotrimeric collagenous peptides clearly underline the advantages of such synthetic constructs over the natural collagen and gelatin or related fragments, for studying the mechanism of action of MMPs on the extracellular matrix. They also open a new strategy of synthesis of biodegradable mimics of collagen.

Acknowledgements

The study was supported by the Sonderforschungsbereich 469 (grant A2) of the Ludwig-Maximilians-University of Munich. The authors would like to thank Prof. K. Kühn, Max-Planck-Institute of Biochemistry, Martinsried, for helpful discussions and Bioselect Nigu for financial support.

References

1. Privalov, P. L. *Adv. Prot. Chem.* **1982**, *35*, 1-104.
2. Baum, J.; Brodsky, B. *Fold. Des.* **1987**, *2*, R53-R60.
3. Beck, K.; Brodsky, B. *J. Struct. Biol.* **1998**, *122*, 17-29.
4. Prockop, D. J.; Kivirikko, K. I. *Annu. Rev. Biochem.* **1995**, *64*, 403-434.
5. Engel, J.; Prockop, D. J. *Annu. Rev. Biophys. Chem.* **1991**, *20*, 137-152.
6. Beck, K.; Boswell, B. A.; Ridgway, C. C.; Bächinger, H. P. *J. Biol. Chem.* **1996**, *271*, 21566-21578.
7. Bulleid, N. J.; Dalley, J. A.; Lees, J. F. *EMBO J.* **1997**, *16*, 6694-6701.
8. Fields, G. B.; Prockop, D. J. *Biopolymers (Pept. Sci.)* **1996**, *40*, 345-357.
9. Uitto, J.; Prockop, D. J. *Biochim. Biophys. Acta* **1974**, *336*, 234-251.
10. Bächinger, H. P.; Davis, J. M. *Int. J. Biol. Macromol.* **1991**, *13*, 152-156.
11. Mayo, K.H. *Biopolymers (Pept. Sci.)* **1996**, *40*, 359-370.
12. Engel, J. *Adv. in Meat Res.* **1987**, *4*, 145-161.
13. Bächinger, H. P.; Bruckner, P.; Timpl, R.; Prockop, D. J.; Engel, J. *Eur. J. Biochem.* **1980**, *106*, 619-632.
14. Goodman, M.; Feng, Y.; Melacini, G.; Taulane, J. P. *J. Am. Chem. Soc.* **1996**, *118*, 5156-5157.
15. Goodman, M.; Melacini, G.; Feng, Y. *J. Am. Chem. Soc.* **1996**, *118*, 10928-10929.
16. Feng, Y.; Melacini, G.; Taulane, J. P.; Goodman, M. *J. Am. Chem. Soc.* **1996**, *118*, 10351-10358.
17. Roth, W.; Heppenheimer, K.; Heidemann, E. R. *Makromol. Chem.* **1979**, *180*, 905-917.
18. Roth, W.; Heidemann, E. R. *Biopolymers* **1980**, *19*, 1909-1917.
19. Fields, C. G.; Lovdahl, C. M.; Miles, A. J.; Yu, Y. C.; Hagen, V. L. M.; Fields, G. B. *Biopolymers* **1993**, *33*, 1695-1707.

116

20. Tanaka, T.; Wada, Y.; Nakamura, H.; Doi, T.; Imanishi, T.; Kodama, T. *FEBS Lett.* **1993**, 334, 272-276.
21. Hoyo, H.; Akamatsu, Y.; Yamauchi, K.; Kinoshita, M. *Tetrahedron* **1997**, 53, 14263-14274.
22. Tanaka, Y.; Kazou, S.; Tanaka, T. *J. Pept. Res.* **1998**, 51, 413-419.
23. Nagase, H.; Woessner, J. F. Jr. *J. Biol. Chem.* **1999**, 274, 21491-21494.
24. Nagase, H.; Fields, G. B. *Biopolymers (Pept. Sci.)* **1996**, 40, 399-416.
25. Fields, C. G.; Grab, B.; Lauer, J. L.; Miles, A. J.; Yu, Y.-C.; Fields, G. B. *Lett. Peptide Sci.* **1996**, 3, 3-16.
26. Fraser, R. D. B.; MacRae, T. P.; Suzuki, E. *J. Mol. Biol.* **1979**, 129, 463-481
27. Bella, J.; Eaton, M.; Brodsky, B.; Berman, H. M. *Science* **1994**, 266, 75-81.
28. Moroder, L.; Battistutta, R.; Besse, D.; Ottl, J.; Pegoraro, S.; Siedler, F. In *Peptides 1996;* Ramage, R.; Epton, R., Eds.; Mayflower Scientific Ltd.: Kingswinford, 1998; pp 59-62.
29. Ottl, J.; Moroder, L. *Tetrahedron Lett.* **1999**, 40, 1487-1490.
30. Ottl, J.; Batistuta, R.; Pieper, M.; Tschesche, H.; Bode, W.; Kühn, K.; Moroder, L. *FEBS Lett.* **1996**, 398, 31-36.
31. Ottl, J.; Moroder, L. *J. Am. Chem. Soc.* **1999**, 121, 653-661.
32. Müller, J. C. D.; Ottl, J.; Moroder, L. *Biochemistry* **2000**, 39, 5111-5116.
33. Brown, F. R. III; Carver, J. P.; Blout, E. R. *J. Mol. Biol.* **1969**, 39, 307-313.
34. Goodman, M.; Bhumralkar, M.; Jefferson, E. A.; Kwak, J.; Locardi, E. *Biopolymers (Pept. Sci.)* **1998**, 47, 127-142.
35. Hofmann, H.; Fietzek, P. P.; Kühn, K. *J. Mol. Biol.* **1978**, 125, 137-165.
36. Engel, J. *Adv. Meat Res.* **1987**, 4, 145-161.
37. Netzel-Arnett, S.; Salari, A.; Goli, U. B.; Van Wart, H. *Ann. New York Acad. Sci.* **1993**, 732, 22-30.
38. Ottl, J.; Gabriel, D.; Murphy, G.; Knäuper, V.; Tominaga, Y.; Nagase, H.; Kröger, M.; Tschesche, H.; Bode, W.; Moroder, L. *Chem. Biol.* **2000**, 7, 119-132.
39. Coussens, L. M.; Werb, Z. *Chem. Biol.* **1996**, 3, 895-904.
40. Berndt, P.; Fields, G. B.; Tirrell, M. *J. Am. Chem. Soc.* **1995**, 117, 9515-9522.

Chapter 9

Peptide-Amphiphile Induction of α-Helical and Triple-Helical Structures

Gregg B. Fields[1,2], Pilar Forns[3], Katarzyna Pisarewicz[1], and Janelle L. Lauer-Fields[1,2]

[1]Department of Chemistry and Biochemistry and [2]Center for Molecular Biology and Biotechnology, Florida Atlantic University, Boca Raton, FL 33431
[3]Laboratori Quimica Organica, Facultat de Farmacia, Universitat de Barcelona, 08028 Barcelona, Spain

Protein-like molecular architecture has often been created by utilizing the ability of peptides to self-assemble and form higher order three-dimensional structures. We have attached pseudo-lipids onto N^{α}-amino groups of peptide chains to create "peptide-amphiphiles." The alignment of amphiphilic compounds at the lipid-solvent interface is used to facilitate peptide alignment and structure initiation and propagation. CD and NMR spectroscopies have been used to examine the secondary or super-secondary structures of a series of peptides both with and without lipophilic hydrocarbon "tails." Overall, the tails (a) do not disrupt the structures of the peptide "head groups," but in fact enhance structure thermal stability and (b) significantly reduce the necessary length for a peptide to have predominantly an α-helical or triple-helical structure in solution. The extent of peptide-amphiphile aggregation appears to be correlated to hydrocarbon tail length. The peptide-amphiphiles described here provide a simple approach for building stable protein structural motifs using peptide head groups, and have potential as therapeutics and for improving biomaterial biocompatibility.

The creation of small, thermally stable, distinctly folded structures is important for both the further understanding of folding principles and the development of uniquely targeted therapeutic agents. There are several approaches by which structures with a predisposition to fold might be induced or stabilized. Templates have been covalently attached to enhance folding (1-3). Introduction of disulfide bridges can stabilize folded structures (4,5) by destabilizing unfolded ones. Non-covalent association of segments has also been used to create folded biomolecules. Such non-covalent assemblies have utilized (a) chelators attached to peptides and appropriate metals to initiate interstrand association or (b) lipids covalently attached to peptides that promote associate *via* hydrophobic interactions (6).

Initial studies from our laboratory examined the induction of protein-like structures by adding mono- and dialkyl chains to a collagen-like peptide (7-12). The alkyl chain tails were found to exert a significant influence on the formation and stabilization of a triple-helical, collagen-like conformation that promotes melanoma cell adhesion and spreading (8-11,13). The induction of α-helical structure by peptide-amphiphiles has not been explored, but would greatly enhance the applicability of these constructs.

We have presently examined peptide-amphiphiles for formation and stabilization of α-helices and collagen-like triple-helices (Figure 1). A previously described model system for α-helices has been chosen in which a repeating peptide heptad sequence, (Glu-Ile-Glu-Ala-Leu-Lys-Ala)$_n$, dimerizes to form a distinct structure at a chain length of 23 residues (14,15). A 16 residue peptide variant was studied to determine if the addition of monoalkyl moeities will promote an α-helical conformation. The structural stability was then compared to peptides and peptide-amphiphiles containing a 39 residue sequence derived from type IV collagen (11,12). Peptide and peptide-amphiphile conformations have been studied by circular dichroism (CD) and one- and two-dimensional NMR spectroscopic techniques. The nature and extent of peptide-amphiphile aggregation has also been examined. Overall, we are trying to determine if alkyl "tails" can be used as a general template for induction of protein-like secondary and tertiary structures and also for interaction with biomaterial surfaces (9,10).

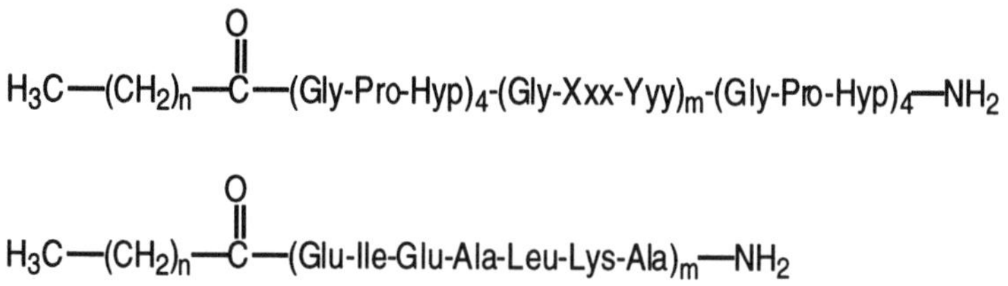

$$H_3C-(CH_2)_n-\overset{O}{\overset{\|}{C}}-(Gly\text{-}Pro\text{-}Hyp)_4\text{-}(Gly\text{-}Xxx\text{-}Yyy)_m\text{-}(Gly\text{-}Pro\text{-}Hyp)_4-NH_2$$

$$H_3C-(CH_2)_n-\overset{O}{\overset{\|}{C}}-(Glu\text{-}Ile\text{-}Glu\text{-}Ala\text{-}Leu\text{-}Lys\text{-}Ala)_m-NH_2$$

Figure 1. General structures of (top) collagen-like triple-helical and (bottom) α-helical monoalkyl peptide-amphiphiles.

Synthesis and Characterization of Peptide-Amphiphiles

The syntheses of α-helical and triple-helical peptide and peptide-amphiphile C-terminal amides were as described previously (8,11,16) using Fmoc solid-phase methodology on either a Perkin Elmer/ABD 431A or 433A Peptide Synthesizer. Peptide-resins were characterized by "embedded" (non-covalent) Edman degradation sequence analysis (17). Peptide-resins were then either (a) cleaved or (b) acylated with the appropriate alkyl tail (11) and then cleaved. Cleavage and side-chain deprotection of peptide-resins and peptide-amphiphile-resins proceeded for 2 h using H₂O–trifluoroacetic acid (1:19). Peptide and peptide-amphiphile cleavage solutions were extracted with methyl tBu ether prior to purification.

Preparative reversed-phase high performance liquid chromatographic purification was performed as described (11,18) on a Rainin AutoPrep System. Matrix-assisted laser desorption/ionization mass spectrometry was performed on a Hewlett-Packard G2025A laser desorption time-of-flight mass spectrometer using a sinapinic acid matrix. α-Helical peptide and peptide-amphiphile mass values are given in Table I. Triple-helical peptide and peptide-amphiphile mass values have been reported (11).

Biophysical Characterization of α-Helical Peptide-Amphiphiles

The present work has examined simple self-assembly systems that may form thermally-stable protein-like molecular architecture. We initially determined if hydrocarbon chains disrupted α-helical secondary structure. CD spectra were recorded over the range λ = 190-250 nm on a JASCO J-600 using a 0.1 cm path-length quartz cell. The CD spectrum of peptide 23r (Table I) in aqueous solution was typical of an α-helix, as two minima are seen at λ = 208 and 222 nm (19). Modification of peptide 23r by C_{10}, C_{12}, C_{16}, or $(C_{12})_2$ lipid tails resulted in peptide-amphiphiles with CD characteristics typical of α-helices (19). The peptide-amphiphiles exhibit CD spectra almost identical to the peptide alone, but with slightly more negative molar ellipticities. The minimum at λ = 208 nm remains slightly more negative than the minimum at λ = 222 nm. Thus, the alkyl chains do not disrupt the secondary structure of 23r. C_{12}-23r retains a high percentage of α-helical structure at 80°C (data not shown), while the peptide 23r has a T_m of 52°C (15). This result is consistant with those obtained for collagen-like peptide-amphiphiles, where triple-helicity was further stabilized against thermal denaturation by alkyl chains (8,11).

Table I. Sequences and [M+H]$^+$ Values for α-Helical Peptides and Peptide-Amphiphiles

Code	Sequence	Theory [M+H]$^+$	Actual [M+H]$^+$
9r	KAEIEALKA-NH$_2$	971.14	971.6
C$_{12}$-9r	CH$_3$(CH$_2$)$_{10}$CO-KAEIEALKA-NH$_2$	1153.46	1152.5
C$_{16}$-9r	CH$_3$(CH$_2$)$_{14}$CO-KAEIEALKA-NH$_2$	1209.57	1208.8
(C$_{12}$)$_2$-9r	C$_{12}$H$_{25}$OCO[C$_{12}$H$_{25}$OCO(CH$_2$)$_2$]-CHNHCO(CH$_2$)$_2$CO-KAEIEALKA-NH$_2$	1536.99	1535.7
12r	EALKAEIEALKA-NH$_2$	1284.50	1283.6
C$_{10}$-12r	CH$_3$(CH$_2$)$_8$CO-EALKAEIEALKA-NH$_2$	1438.77	1435.5
C$_{12}$-12r	CH$_3$(CH$_2$)$_{10}$CO-EALKAEIEALKA-NH$_2$	1466.82	1466.3
C$_{16}$-12r	CH$_3$(CH$_2$)$_{14}$CO-EALKAEIEALKA-NH$_2$	1522.93	1522.0
(C$_{12}$)$_2$-12r	C$_{12}$H$_{25}$OCO[C$_{12}$H$_{25}$OCO(CH$_2$)$_2$]CHNHCO-(CH$_2$)$_2$CO-EALKAEIEALKA-NH$_2$	1850.35	1848.4
16r	KA[EIEALKA]$_2$-NH$_2$	1726.10	1725.5
C$_6$-16r	CH$_3$(CH$_2$)$_4$CO-KA[EIEALKA]$_2$-NH$_2$	1824.26	1820.3
C$_{10}$-16r	CH$_3$(CH$_2$)$_8$CO-KA[EIEALKA]$_2$-NH$_2$	1880.37	1878.6
C$_{12}$-16r	CH$_3$(CH$_2$)$_{10}$CO-KA[EIEALKA]$_2$-NH$_2$	1908.32	1906.2
C$_{16}$-16r	CH$_3$(CH$_2$)$_{14}$CO-KA[EIEALKA]$_2$-NH$_2$	1964.53	1961.3
(C$_{12}$)$_2$-16r	C$_{12}$H$_{25}$OCO[C$_{12}$H$_{25}$OCO(CH$_2$)$_2$]CHNHCO-(CH$_2$)$_2$CO-KA[EIEALKA]$_2$-NH$_2$	2291.95	2289.5
23r	KA[EIEALKA]$_3$-NH$_2$	2480.70	2480.6
C$_{10}$-23r	CH$_3$(CH$_2$)$_8$CO-KA[EIEALKA]$_3$-NH$_2$	2634.97	2632.4
C$_{12}$-23r	CH$_3$(CH$_2$)$_{10}$CO-KA[EIEALKA]$_3$-NH$_2$	2663.02	2659.9
C$_{16}$-23r	CH$_3$(CH$_2$)$_{14}$CO-KA[EIEALKA]$_3$-NH$_2$	2719.13	2718.0
(C$_{12}$)$_2$-23r	C$_{12}$H$_{25}$OCO[C$_{12}$H$_{25}$OCO(CH$_2$)$_2$]CHNHCO-(CH$_2$)$_2$CO-KA[EIEALKA]$_3$-NH$_2$	3046.55	3046.0

The minimum length necessary for a peptide-amphiphile to exhibit an α-helical structure was next examined. Peptides of length 16, 12, and 9 amino acids, designated 16r, 12r, and 9r, respectively (Table I), were studied. Peptide 16r does not exhibit an α-helical CD spectrum in aqueous solution (19), but rather shows features of both turns and random coils. In contrast, peptide-amphiphiles C$_{10}$-16r, C$_{16}$-16r, and (C$_{12}$)$_2$-16r have CD spectra consistant with α-helical structure (19). (C$_{12}$)$_2$-16r shows the highest content of α-helicity based on the relative molar ellipticity values at λ = 208 and 222 nm. The minimum at λ = 208 nm remains slightly more negative than the minimum at λ = 222 nm for C$_{10}$-16r, C$_{16}$-16r, and (C$_{12}$)$_2$-16r.

Peptide 12r does not exhibit an α-helical CD spectrum in aqueous solution, nor does the C$_{10}$-12r peptide amphiphile (19). Peptide-amphiphiles C$_{16}$-12r and (C$_{12}$)$_2$-12r exhibit CD spectra typical of α-helices (19), although the two minima molar ellipticities are less negative than that observed for the 16r peptide-amphiphiles.

The smallest peptide tested was the 9 residue 9r (Table I). The CD spectrum of 9r showed features typical of random coils, with no α-helical structure present (data not shown). Furthermore, none of the 9r peptide-amphiphiles had an α-helical-like structure in solution (data not shown).

Based on the initial CD studies, 16r showed the most distinct structural differences between the peptide and peptide-amphiphiles. Thus, further structural characterization experiments were performed with peptide 16r and related peptide-amphiphiles. The importance of covalent attachment of the hydrocarbon chain was examined by comparing CD spectra of 16r in the presence and absence of exogenous hydrocarbon chains. The spectrum was identical for 16r in the presence of C_{10} or C_{16} or in the absence of these hydrocarbon chains (data not shown).

The structural characteristics of 16r and 16r peptide-amphiphiles in solvent mixtures were compared. TFE has been used to induce α-helical structure in peptide 23r (15). The effect of TFE on the stuctures of 16r, C_6-16r, and C_{16}-16r was thus examined. The effect of either 25% or 50% TFE on the 16r peptide was significant. Changes in the CD spectra indicated the formation of α-helical structure upon TFE addition (Table II). The CD spectrum of C_6-16r in aqueous solution at 5°C was consistant with a moderate α-helical content (Table II). The values for $[\Theta]_{208}$ and $[\Theta]_{222}$ were approximately equivalent. Addition of TFE to a final concentration of 50% resulted in futher induction of α-helical structure, and a decrease in the $[\Theta]_{222}/[\Theta]_{208}$ ratio (18). The CD spectrum of C_{16}-16r in aqueous solution at 5°C was consistant with high α-helical content (Table II). The $[\Theta]_{208}$ value was slightly more negative than $[\Theta]_{222}$ (18). Addition of TFE minimally enhanced α-helicity (Table II), and resulted in a decrease in the $[\Theta]_{222}/[\Theta]_{208}$ ratio (18). A decrease in this ratio has been correlated with a transition from dimeric α-helical coiled-coils to monomeric α-helices (15). TFE appears to be acting on the C_6-16r peptide-amphiphile by weakening the hydrophobic interactions of the alkyl chains (causing an oligomer to monomer transition) while enhancing the local α-helical propensities in the head group.

The relative stabilities of α-helical structure within the 16r peptide-amphiphiles were compared. Thermal transition curves were obtained by recording $[\Theta]_{222}$ while the temperature was continuously increased in the range of 25-80°C at a rate of 0.2°C/min. Temperature was controlled using a JASCO PTC-348WI temperature control unit. For samples exhibiting sigmoidal melting curves, the melting temperature (T_m) was evaluated from the midpoint of the transition. The thermal transition from α-helix to "random coil" state for C_6-16r, C_{16}-16r, and $(C_{12})_2$-16r indicated T_m values of 46, 65, and 59°C, respectively (Table III). For the 16r peptide, the longer the monoalkyl tail the greater the α-helical thermal stability, as has been seen for triple-helical head groups (11). The transition for C_6-16r was somewhat broad, while C_{16}-16r exhibited a more sigmoidal transition. Prior studies on dimeric α-helical coiled coil model peptides showed that as peptide chain length increased the melting temperature increased and denaturation curves became broader (14,20). It has been proposed that the broad transition may be the result of a non-cooperative melt of the α-helices, i.e. some residues retain substantial α-helical character while others approach a random coil conformation (14). C_6-16r may exhibit a broader melting transition than C_{16}-16r due to the more extensive and stable aggregate formed by the C_{16} tail. Increasing temperature does increase the strength of the hydrophobic effect and promote aggregation (21), and thus the C_6-16r peptide-amphiphile may first be "driven" into a more aggregated form with increasing temperature before finally denaturing. Also, the α-helical head group may melt somewhat non-cooperatively.

Table II. Helical Content of Peptide 16r and 16r Peptide-Amphiphiles

Peptide or Peptide-Amphiphile	Solvent	$[\Theta]_{222}$ ($°cm^2/dmol$)	HC^a (%)
16r	H_2O	-2432.3	13
16r	TFE–H_2O (1:1)	-16157.5	84
C_6-16r	H_2O	-11053.5	58
C_6-16r	TFE–H_2O (1:1)	-19714.3	100
C_{16}-16r	H_2O	-17034.9	89
C_{16}-16r	TFE–H_2O (1:1)	-19131.5	100

[a]HC = α-helical content at ~0.25 mM and 25°C, relative to C_{16}-16r in TFE–H_2O (1:1) at 5°C.

Table III. T_m Values for α-Helix Or Triple Helix ⇔ Coil Transitions

Peptide or Peptide-Amphiphile[a]	T_m (°C)
Lys-Ala-[Glu-Ile-Glu-Ala-Leu-Lys-Ala]$_2$-NH$_2$	NH[b]
C_6-Lys-Ala-[Glu-Ile-Glu-Ala-Leu-Lys-Ala]$_2$-NH$_2$	46.0
C_{16}-Lys-Ala-[Glu-Ile-Glu-Ala-Leu-Lys-Ala]$_2$-NH$_2$	65.0
$(C_{12})_2$-Glu-C_2-Lys-Ala-[Glu-Ile-Glu-Ala-Leu-Lys-Ala]$_2$-NH$_2$	59.0
(Gly-Pro-Hyp)$_4$-[IV-H1]-(Gly-Pro-Hyp)$_4$-NH$_2$	35.6
C_6-(Gly-Pro-Hyp)$_4$-[IV-H1]-(Gly-Pro-Hyp)$_4$-NH$_2$	42.2
C_8-(Gly-Pro-Hyp)$_4$-[IV-H1]-(Gly-Pro-Hyp)$_4$-NH$_2$	45.6
C_{10}-(Gly-Pro-Hyp)$_4$-[IV-H1]-(Gly-Pro-Hyp)$_4$-NH$_2$	51.3
C_{12}-(Gly-Pro-Hyp)$_4$-[IV-H1]-(Gly-Pro-Hyp)$_4$-NH$_2$	55.0
C_{14}-(Gly-Pro-Hyp)$_4$-[IV-H1]-(Gly-Pro-Hyp)$_4$-NH$_2$	63.1
C_{16}-(Gly-Pro-Hyp)$_4$-[IV-H1]-(Gly-Pro-Hyp)$_4$-NH$_2$	69.8
$(C_{12})_2$-Glu-C_2-(Gly-Pro-Hyp)$_4$-[IV-H1]-(Gly-Pro-Hyp)$_4$-NH$_2$	71.2

[a][IV-H1] = Gly-Val-Lys-Gly-Asp-Lys-Gly-Asn-Pro-Gly-Trp-Pro-Gly-Ala-Pro.
[b]NH = not helical.

NMR spectroscopy was used to further characterize the structures of 16r, C_6-16r, and C_{16}-16r. NMR spectra were acquired on a 500 MHz Varian Inova spectrometer at room temperature. Freeze-dried samples were dissolved in D_2O–H_2O (1:9) or TFE-d_3–H_2O (1:1) at concentrations of 0.5-3 mM, depending upon the solubility of the peptides and peptide-amphiphiles. ^1H chemical shifts were expressed relative to sodium 3-(trimethylsilyl)tetradeuteriopriopionate. One of the ^1H-NMR spectral parameters indicative of α-helices is the dispersion and resolution of NH resonances upon transition from the random coil state (22). The one-dimensional NMR spectrum of 16r shows a dispersion from 8.00-8.74 ppm and many overlapping peaks (18). Addition of TFE to a final concentration of 50%, which substantially increases the α-helical content of 16r based on CD measurements (Table II), results in a dispersion from 7.70-8.68 ppm and improved resolution of peaks (18). The ^1H-NMR spectrum of C_6-16r in D_2O–H_2O (1:9) shows a dispersion from 7.76-8.60 ppm (18), which is similar to 16r in TFE–H_2O (1:1). A similar dispersion is seen in the ^1H-NMR spectrum of C_{16}-16r in D_2O–H_2O (1:9) (data not shown). When TFE was added to C_6-16r, better resolution of the NH region was seen, but only slightly better dispersion (7.73-8.61 ppm) (18). This result is consistent with CD measurements, where TFE modestly enhances the α-helical content of C_6-16r (Table II).

The ^1H-NMR resonances for C_6-16r in D_2O–H_2O (1:9) at pH ~4 and 25°C were assigned (18). The αCH proton resonances are upfield shifted, consistent with α-helical conformation (14,22). An average α proton shift of -0.39 ppm is seen when an amino acid is transferred from a random coil to an α-helical environment (22). The $\Delta\delta_{\alpha CH}$ values can be calculated for the αCH protons in C_6-16r by subtracting the random coil αCH chemical shift for a specific residue (22) from the observed αCH chemical shift for that residue (18). All $\Delta\delta_{\alpha CH}$ values are ≤ 0, suggesting that the 16r peptide region is α-helical (23). The $\Delta\delta_{\alpha CH}$ values change periodically along the peptide chain with a 3-4 repeat pattern, with five maximum negative $\Delta\delta_{\alpha CH}$ values at positions 1, 4, 8, 11, and 14 and five maximum positive $\Delta\delta_{\alpha CH}$ values at positions 2, 6, 9, 13, and 16. These results are consistent with those reported by Zhou et al. (23), in that the α protons within an α-helix showed a 3-4 repeat periodicity in chemical shifts.

To obtain two-dimensional NMR spectra, total correlation spectroscopy (TOCSY) and nuclear Overhauser effect spectroscopy (NOESY) were performed with 512 t1 increment and 1024 complex data points in the t2 dimension. TOCSY spectra were obtained at mixing times of 40-150 msec. NOESY spectra were obtained at mixing times of 250 msec. The spectral widths were 5500 Hz in both dimensions. The NOESY spectra of an α-helix should show an increase in the ^1H-^1H connectivities in the NH-NH and in the NH-CαH regions compared to the random coil. More specifically, one should see strong $d_{NN}(i,i+1)$ connectivities and moderate $d_{NN}(i,i+2)$ and $d_{\alpha N}(i,i+3)$ connectivities (14,24-26). The NOESY spectra of 16r in D_2O–H_2O (1:9) show little or no α-helical structure based upon the number of cross peaks in the NH-NH and NH-CαH regions (*18*). The NOESY spectra of C_6-16r in D_2O–H_2O (1:9) show more cross peaks in the NH-NH and NH-CαH regions than peptide 16r (*18*). One could identify a large number of ^1H-^1H NOE connectivities in the NH-CαH region, including $d_{\alpha N}(i,i+3)$ for residues 1 and 4, 2 and 5, 3 and 6, 4 and 7, 5 and 8, 6 and 9, 7 and 10, 8 and 11, 9 and 12, 10 and 13, 12 and 15, and 13 and 16 in C_6-16r (*18*).

Several criteria based on CD and NMR spectroscopic results have allowed us to classify the head group structure of the aforementioned peptide-amphiphiles as primarily α-helical. Specifically, (a) CD spectra characteristic of α-helices, (b) dispersion of the ^1H-NMR chemical shifts in the NH region, (c) relative ^1H-NMR chemical shift values for αCH proton resonances, and (d) NOESY spectral $d_{\alpha N}(i,i+3)$ connectivities have been used to designate α-helices. Based on these criteria, a peptide containing 16 amino acid residues may form an α-helix in combination with a C_6, C_{10}, or C_{16} monoalkyl or $(C_{12})_2$ dialkyl chain, while the peptide alone has little α-helical structure. In aqueous solution, the peptide alone requires a length of 23 amino acids to form primarily an α-helical dimer (*15*). The effects of the monoalkyl tails are not simply to eliminate charges that lead to unfavorable interactions with the α-helix dipole, as (a) peptide 16r with an *N*-terminal acetyl group has little α-helical structure (*15*) and (b) the thermal stability of the α-helix increases with increasing tail length (Table III).

Biophysical Comparison of α-Helical and Triple-Helical Peptide-Amphiphiles

To investigate the versatility of peptide-amphiphiles, we have compared the effect of lipidation on stabilizing α-helical and triple-helical structures. A series of monoalkyl tails have been coupled to a collagen-derived sequence (Gly-Pro-Hyp)₄-Gly-Val-Lys-Gly-Asp-Lys-Gly-Asn-Pro-Gly-Trp-Pro-Gly-Ala-Pro-(Gly-Pro-Hyp)₄ [designated (Gly-Pro-Hyp)₄-[IV-H1]-(Gly-Pro-Hyp)₄] (*10*). T_m values were found to increase with monoalkyl tail chain length (Table III). Thus, both monoalkyl and dialkyl hydrocarbon chains could be used to induce and enhance the thermal stability of triple-helical and α-helical head groups.

Heteronuclear single quantum coherence (HSQC) and inverse-detected ^1H-^{15}N NMR spectroscopy was used to examine the structure and dynamics of (Gly-Pro-Hyp)₄-[IV-H1]-(Gly-Pro-Hyp)₄, and C_6-(Gly-Pro-Hyp)₄-[IV-H1]-(Gly-Pro-Hyp)₄ (*11*). Overall, the peptide head group of the C_6-(Gly-Pro-Hyp)₄-[IV-H1]-(Gly-Pro-Hyp)₄ peptide-amphiphile appeared to form a continuous triple-helix (*12*). Our two-dimensional NMR studies of C_6-16r indicated that a continuous α-helix was formed within the peptide-amphiphile head group. Thus, the helical structures induced by the alkyl chains (whether α-helical or triple-helical) are well-ordered and continuous.

Aggregation of α-Helical and Triple-Helical Peptide-Amphiphiles

To evaluate the size of potential peptide-amphiphile aggregates, C_6-16r and C_{16}-16r were subjected to size-exclusion chromatography (SEC) and native gel electrophoresis (GE) (18). For SEC experiments, peptide 16r was resynthesized to include an N-terminal Tyr as a chromophore. C_{16}-Tyr-16r and C_{16}-16r showed identical structures by CD analysis, as did C_6-Tyr-16r and C_6-16r (data not shown). SEC was performed initially using a 1.5 x 190 cm column of Bio-GelTM A-0.5m Gel. The eluent was monitored by measuring the absorbance at λ = 280 nm. Protein standards (1-5 mg) or peptide-amphiphiles (~1.5 mM) were applied to the column in 0.5 mL of the phosphate buffer, pH 7.0, and eluted with the same buffer at a flow rate of 0.3 mL/min. SEC was repeated using 85 ml bed volumes of Sephadex G-50 or G-200 in 0.25 M phosphate buffer containing 0.5 M NaCl. Samples (3 mg/ml in 1 ml) were eluted at 4.0 and 1.0 ml/min from the G-50 and G-200 columns, respectively. Apparent molecular weights were determined by interpolation from the standard curves. SEC of C_{16}-Tyr-16r resulted in an apparent molecular weight of ~130 kDa (Table IV). C_6-Tyr-16r had a molecular weight <12 kDa (Table IV). Native GE was performed for the peptide-amphiphiles for comparison to the SEC results. Peptide-amphiphiles were dissolved in sample buffer containing 1.0 M Tris•HCl, 15% glycerol, and 0.1 mg/mL bromophenol blue. Samples were electrophoresed using a 4-15% Tris•HCl native gel in buffer containing 25 mM Tris•HCl and 0.19 M glycine. Sample concentrations were 0.75-2.0 mM. Gels were run for 3.25 h at 20 mA. C_6-16r showed no substantial aggregation under both non-denaturing and denaturing conditions (data not shown). Under non-denaturing conditions, C_{16}-Tyr-16r migrated as two distinct species, the first sharp band with a molecular mass of ~130 kDa and the second diffuse band with a molecular mass <75 kDa (data not shown). Mixing C_{16}-Tyr-16r with TFE (to a final TFE concentration of 50%) resulted in the disappearance of the ~130 kDa band and a slightly higher molecular mass for the diffuse band (data not shown). Both SEC and GE indicated that the C_{16} monoalkyl tail associates to form moderate size aggregates, while the C_6 monoalkyl tail may associate to form only small aggregates.

SEC experiments were continued for the monoalkyl triple-helical peptide-amphiphiles. Peptide-amphiphiles of the sequence C_n-(Gly-Pro-Hyp)$_4$-[IV-H1]-(Gly-Pro-Hyp)$_4$, where n = 6, 8, 10, 12, 14, or 16, were found to form aggregates of size that directly correlated to alkyl tail length (Table IV). The shortest alkyl chain, C_6, resulted in two distinct sizes of peptide-amphiphile aggregates. The smaller aggregate was 28.6 kDa, while the larger was 165.3 kDa. For each of the other alkyl chain lengths, one aggregate of a distinct molecular mass was found. When one compares the aggregates formed by the α-helical peptide-amphiphiles with those formed by the triple-helical peptide-amphiphiles, the number of *monomeric* peptide-amphiphile chains per aggregate is similar for either an α-helical or triple-helical peptide-amphiphile of the same alkyl chain length. For example, a C_{16} alkyl chain results in aggregation of 66 single α-helical peptide-amphiphiles or 22 trimeric triple-helical peptide-amphiphiles. In similar fashion, a C_{10} alkyl chain results in aggregation of 50-51 single α-helical peptide-amphiphiles (18) or 18 trimeric triple-helical peptide-amphiphiles. Thus, the size of the aggregate appears to be dependent upon alkyl chain length, regardless of peptide head group structure.

Table IV. Aggregation of Monoalkyl Peptide-Amphiphiles

Peptide-Amphiphile[a]	MW (kDa)	
	Monomer[b]	Actual
α-Helices		
C_6-Tyr-Lys-Ala-[Glu-Ile-Glu-Ala-Leu-Lys-Ala]$_2$-NH$_2$	1.82	<12.0
C_{10}-Tyr-Lys-Ala-[Glu-Ile-Glu-Ala-Leu-Lys-Ala]$_2$-NH$_2$	1.88	95.0
C_{16}-Tyr-Lys-Ala-[Glu-Ile-Glu-Ala-Leu-Lys-Ala]$_2$-NH$_2$	1.96	130.0
Triple-Helices		
C_6-(Gly-Pro-Hyp)$_4$-[IV-H1]-(Gly-Pro-Hyp)$_4$-NH$_2$	11.0	28.6
		165.3
C_8-(Gly-Pro-Hyp)$_4$-[IV-H1]-(Gly-Pro-Hyp)$_4$-NH$_2$	11.1	195.4
C_{10}-(Gly-Pro-Hyp)$_4$-[IV-H1]-(Gly-Pro-Hyp)$_4$-NH$_2$	11.2	202.4
C_{12}-(Gly-Pro-Hyp)$_4$-[IV-H1]-(Gly-Pro-Hyp)$_4$-NH$_2$	11.3	202.4
C_{14}-(Gly-Pro-Hyp)$_4$-[IV-H1]-(Gly-Pro-Hyp)$_4$-NH$_2$	11.4	214.6
C_{16}-(Gly-Pro-Hyp)$_4$-[IV-H1]-(Gly-Pro-Hyp)$_4$-NH$_2$	11.4	251.6

[a][IV-H1] = Gly-Val-Lys-Gly-Asp-Lys-Gly-Asn-Pro-Gly-Trp-Pro-Gly-Ala-Pro.
[b]For triple-helical peptide-amphiphiles, a monomer is composed of three peptide strands intertwined.

Fourier transform infrared (FTIR) spectroscopy was used to examine the interaction of peptide-amphiphile tails in a variety of solvents. Initial studies by GE indicated that α-helical and triple-helical peptide amphiphile aggregates were disrupted by TFE but not by DCM. $CHCl_3$ disrupted triple-helical, but not α-helical, peptide-amphiphile aggregates. These three solvents were thus used for FTIR spectroscopic studies. Samples were analyzed on KBr windows between 2000-4000 cm^{-1}. This region encompasses C-H stretching modes for n-alkyl chains (27). In TFE, which disrupts both peptide-amphiphile aggregates, the C-H stretching modes were very similar for the C_{16} acid and the C_{16} peptide-amphiphiles (Table V). In DCM, which does not disrupt either peptide-amphiphile aggregate, the stretching modes for the α-helical peptide-amphiphile are very different than for the C_{16} acid, while the triple-helical peptide-amphiphile has some similar modes and some different ones (Table V). In $CHCl_3$, which only disrupts the triple-helical peptide-amphiphile, the stretching modes are very different between the C_{16} acid and the α-helical peptide-amphiphile and very similar between the C_{16} acid and the triple-helical peptide-amphiphile (Table V). Overall, stretching modes are very similar between the C_{16} acid and the peptide-amphiphiles when aggregation is disrupted, and very different when the aggregates are intact. It appears that peptide-amphiphile aggregation effects both the peptide head group and alkyl tail structures. Also, there are subtle differences in the tail interactions based on the head group structure.

Table V. FTIR Spectroscopic Stretching Modes for the
C_{16} Alkyl Chain and C_{16} Peptide-Amphiphiles

Solvent	Alkyl Tail or Peptide-Amphiphile	C-H Bands (cm^{-1})
TFE	C_{16} acid	3628.39, 3367.09, 2961.70, 2889.77
TFE	C_{16}-α-helical	3684.25, 3365.92, 2959.14, 2886.97
TFE	C_{16}-triple-helical	3687.33, 3367.08, 2955.15, 2887.44
DCM	C_{16} acid	3053.58, 2986.21, 2927.27, 2854.10
DCM	C_{16}-α-helical	3086.11, 2952.69, 2922.39
DCM	C_{16}-triple-helical	3053.55, 2986.19, 2918.12
CHCl₃	C_{16} acid	3018.88, 2926.92, 2854.61, 2399.84
CHCl₃	C_{16}-α-helical	3050.43, 2991.69, 2415.55, 2359.06, 2331.72
CHCl₃	C_{16}-triple-helical	3468.18, 3411.80, 3019.68, 2399.86

Peptide-Amphiphiles As a General Approach for Construction of Molecular Architectures

In the present study we have examined the association of lipophilic compounds to create a surface that aligns peptide strands and induces and/or stabilizes protein-like secondary, supersecondary, and tertiary structures within these peptide sequences. Initial studies demonstrated that dialkyl ester compounds of carbon chain lengths 12-18 stabilize triple-helical conformation (8,9). Subsequently, monoalkyl chains were found to exert a similar stabilizing effect on triple-helical structure (11,12). The potential of peptide-amphiphiles for providing a general "folding surface" was further examined by using structural elements that diverge significantly from triple-helices. α-Helices are an abundant and biologically relevant secondary structure, and thus were chosen for study in the peptide-amphiphile system. Alkyl chains were found to induce α-helical structure in a 16 residue peptide head group. For both α-helical and triple-helical peptide-amphiphiles, a direct correlation was observed between head group thermal stability and alkyl chain length. Thus, not only are different helices induced and stabilized within peptide-amphiphiles, but structural stability and the degree of aggregation can be modulated by alkyl chain length.

Overall, peptide-amphiphiles can be utilized to create a variety of protein-like molecular architectures. Hydrophobic interactions between alkyl tails appear to be the stabilizing force for these peptide-amphiphile systems. In their present form, peptide-amphiphiles show particular promise for modifying biomaterial surfaces. Peptide-amphiphile interaction with surfaces can proceed by direct interaction on hydrophobic surfaces or mixed bilayer formation on hydrophilic surfaces (9). Once layered on surfaces, the peptide head group appears to be directed upward (7,9), allowing for interaction with the biological environment. Surface-coated peptide-amphiphiles have been shown to promote cellular recognition and signaling (9,10,13,28,29), and thus could be used to improve biomaterial biocompatibility. As the full range of peptide-amphiphile aggregate structures are explored, the use of alkyl tails may eventually play a role in protein design and construction.

Acknowledgments

We thank Dr. Frank Marí for assistance with the NMR experiments. This research is supported by the NIH (HL62427 and AR01929). P.F. was a recipient of a grant for training of postdoctoral staff from Ministerio de Educación y Cultura, Spain, and G.B.F. was a recipient of an NIH RCDA.

References

1. Dawson, P. E.; Kent, S. B. H. *J. Am. Chem. Soc.* **1993**, *115*, 7263-7266.
2. Mutter, M.; Tuchscherer, G. G.; Miller, C.; Altmann, K. H.; Carey, R. I.; Wyss, D. F.; Labhardt, A. M.; Rivier, J. E. *J. Am. Chem. Soc.* **1992**, *114*, 1463-1470.
3. Vuilleumer, S.; Mutter, M. *Biopolymers* **1993**, *33*, 389-400.
4. Zhou, N. E.; Kay, C. M.; Hodges, R. S. *Biochemistry* **1993**, *32*, 3178-3187.
5. Ottl, J.; Moroder, L. *J. Am. Chem. Soc.* **1999**, *121*, 653-661.
6. Borgia, J. A.; Fields, G. B. *Trends Biotech.* **2000**, *18*, 243-251.
7. Berndt, P.; Fields, G. B.; Tirrell, M. *J. Am. Chem. Soc.* **1995**, *117*, 9515-9522.
8. Yu, Y.-C.; Berndt, P.; Tirrell, M.; Fields, G. B. *J. Am. Chem. Soc.* **1996**, *118*, 12515-12520.
9. Yu, Y.-C.; Pakalns, T.; Dori, Y.; McCarthy, J. B.; Tirrell, M.; Fields, G. B. *Meth. Enzymol.* **1997**, *289*, 571-587.
10. Fields, G. B.; Lauer, J. L.; Dori, Y.; Forns, P.; Yu, Y.-C.; Tirrell, M. *Biopolymers* **1998**, *47*, 143-151.
11. Yu, Y.-C.; Tirrell, M.; Fields, G. B. *J. Am. Chem. Soc.* **1998**, *120*, 9979-9987.
12. Yu, Y.-C.; Roontga, V.; Daragan, V. A.; Mayo, K. H.; Tirrell, M.; Fields, G. B. *Biochemistry* **1999**, *38*, 1659-1668.
13. Dori, Y.; Bianco-Peled, H.; Satija, S. K.; Fields, G. B.; McCarthy, J. B.; Tirrell, M. *J. Biomed. Mater. Res.* **2000**, *50*, 75-81.
14. Houston, M. E., Jr.; Campbell, A. P.; Lix, B.; Kay, C. M.; Sykes, B. D.; Hodges, R. S. *Biochemistry* **1996**, *35*, 10041-10050.

15. Su, J. Y.; Hodges, R. S.; Hay, C. M. *Biochemistry* **1994**, *33*, 15501-15510.
16. Lauer, J. L.; Fields, C. G.; Fields, G. B. *Lett. Peptide Sci.* **1995**, *1*, 197-205.
17. Fields, C. G.; VanDrisse, V. L.; Fields, G. B. *Peptide Res.* **1993**, *6*, 39-47.
18. Forns, P.; Lauer-Fields, J. L.; Gao, S.; Fields, G. B. *Biopolymers* **2000**, accepted for publication.
19. Forns, P.; Fields, G. B. *Polymer Preprints* **2000**, *41*, in press.
20. Lau, S. Y. M.; Taneja, A. K.; Hodges, R. S. *J. Biol. Chem.* **1984**, *259*, 13253-13261.
21. Fields, G. B.; Alonso, D. O. V.; Stigter, D.; Dill, K. A. *J. Phys. Chem.* **1992**, *96*, 3974-3981.
22. Wishart, D. S.; Sykes, B. D.; Richards, F. M. *J. Mol. Biol.* **1991**, *222*, 311-333.
23. Zhou, N. E.; Zhu, B.-Y.; Sykes, B. D.; Hodges, R. S. *J. Am. Chem. Soc.* **1992**, *114*, 4320-4326.
24. Wüthrich, K.; Billeter, M.; Braun, W. *J. Mol. Biol.* **1984**, *180*, 715-740.
25. Wagner, G.; Neuhaus, D.; Wirgitter, E.; Vasák, M.; Kägi, J. H. R.; Wüthrich, K. *J. Mol. Biol.* **1986**, *187*, 131-135.
26. Dyson, H. J.; Wright, P. E. *Annu. Rev. Biophys. Chem.* **1991**, *20*, 519-538.
27. Snyder, R. G.; Strauss, H. L.; Elliger, C. A. *J. Phys. Chem.* **1982**, *86*, 5145-5150.
28. Winger, T. M.; Ludovice, P. J.; Chaikof, E. L. *Biomaterials* **1996**, *17*, 437-440.
29. Pakalns, T.; Haverstick, K. L.; Fields, G. B.; McCarthy, J. B.; Mooradian, D. L.; Tirrell, M. *Biomaterials* **1999**, *20*, 2265-2279.

Macromolecular Assemblies

Chapter 10

Synthesis, Helical Chirality, and Self-Assembling Hierarchical Structures of Amino Acid-Containing Polyacetylenes

Ben Zhong Tang[1,*], Kevin K. L. Cheuk[1], Fouad Salhi[1], Bingshi Li[2], Jacky W. Y. Lam[1], John A. K. Cha[1], and Xudong Xiao[2]

Departments of [1]Chemistry and [2]Physics, Hong Kong University of Science and Technology, Clear Water Bay, Kowloon, Hong Kong, China
[*]Corresponding author: e-mail: tangbenz@ust.hk

Creation of unnatural helical macromolecules and construction of biomimetic hierarchical structures are of both scientific value and technological implication. In this work, we synthesized a group of amphiphilic polyacetylenes of high molecular weights (up to million daltons) in high yields. The helical chirality of the macromolecular chains can be continuously and reversibly tuned by simple external stimuli such as solvent and pH. The helical polymers, in response to the changes in their environments, self-associate into macromolecular assemblies reminiscent of natural organizational structures such as double helix, twisted cable, spherical vesicle, hairpin loop, extended fibril, coiling ribbon, honeycomb pattern, and mollusk shape.

Helicity is a structural feature of biomacromolecules and is expressed at all organizational levels of the molecular machinery of living systems, e.g., α helix of proteins, double helix of DNA, triple helix of collagen, and spiral bacterium of *Spirillum* (1, 2). Hierarchy is a characteristic of biological self-assembling processes, as exemplified by protein folding, which involves a sequential set of structural intermediates, each being more organized at a higher structural level than the one before it. Many biological functions performed by biomacromolecules are directly associated with their well-defined hierarchical structures, whose formation is largely regulated by the helical conformation of the polymer chains and by the packing information encoded in their building blocks. Much, however, remains to be learned about the helicity and hierarchy in nature because of the involved complexity.

Development of biomimetic helical polymers is a topic of current interest (*3–17*). Studies of the simple unnatural systems may lead to a better understanding of the complex natural systems, to the possible control of the elementary steps in the natural construction processes, and to the exploration of efficient strategies for assembling molecular components into hierarchical architectures. We here report a group of amphiphilic polyacetylenes that possess helical chirality and are capable of self-assembling. The helicity of the macromolecular chains can be readily tuned and the assembled structures are reminiscent of the shapes and patterns found in living world.

Synthesis of Amino Acid-containing Polyacetylenes

Polyacetylene is an archetypal conjugate polymer of high electrical conductivity (*18*) and its derivatives with appropriate substituents show unique electronic and optical properties such as photoconductivity, liquid crystallinity, and photo- and electroluminescence (*19–21*). Wrapping the molecular wires of the conductive polyacetylene backbones with biocompatible pendants of naturally occurring building blocks such as amino acids may lead to the development of biomaterials for bioinspired technologies such as artificial nerve systems and photosynthesis devices. We thus designed a synthetic route to amino acid-containing polyacetylenes (Scheme 1).

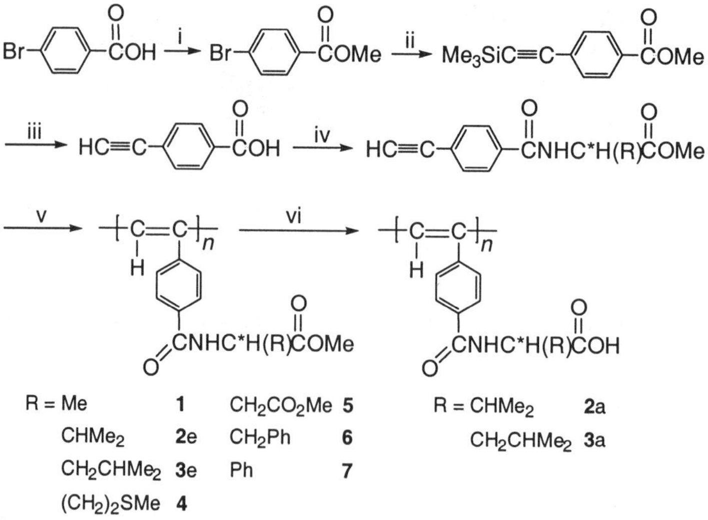

Br
COH
i
Br
COMe
ii
Me3SiC≡C
COMe
iii
HC≡C
COH
iv
HC≡C
CNHC*H(R)COMe
v
C=C
n
H
CNHC*H(R)COMe
vi
C=C
n
H
CNHC*H(R)COH
R = Me 1
CHMe2 2e
CH2CHMe2 3e
(CH2)2SMe 4
CH2CO2Me 5
CH2Ph 6
Ph 7
R = CHMe2 2a
CH2CHMe2 3a

*Scheme 1. Synthesis of amino acid-containing polyacetylenes **1–7**: (i) H_2SO_4, MeOH, reflux, 2 h, 90%; (ii) $Me_3SiC{\equiv}CH$, $(Ph_3P)_2PdCl_2$, CuI, Et_3N, 25 °C, 8 h, 96%; (iii) KOH, MeOH, reflux, 4 h, 84%; (iv) $SOCl_2$, DMF, DCM, 25 °C, and then L-amino acid methyl ester hydrochloride, DCM, pyridine, 25 °C, 8 h, 55–63%; (v) $[Rh(nbd)Cl]_2$, Et_3N, THF, 25 °C, 24 h, 76–97%, M_w 0.17–1.47 MDa; and (vi) KOH, MeOH, 25 °C, 0.5–1 h, 100%, M_w 0.41–1.05 MDa [All the esters (e) can be converted to the corresponding acids (a) by the base-catalyzed hydrolysis, two of which are performed in the present study].*

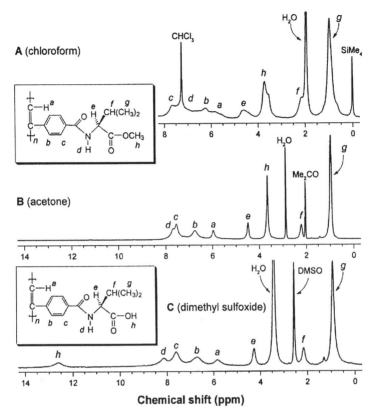

Figure 1. NMR spectra of 2e in (A) chloroform-d and (B) acetone-d₆ and (C) 2a in DMSO-d₆

Esterification of 4-bromobenzoic acid followed by Pd-catalyzed coupling with (trimethylsilyl)acetylene yields methyl [4-(trimethylsilyl)ethynyl]benzoate, hydrolysis of which gives 4-ethynylbenzoic acid (*22*). Condensation of the acid with L-amino acid methyl esters results in the formation of the *p*-substituted phenylacetylene monomers, whose polymerizations are readily initiated by a rhodium catalyst (*23*). Selective hydrolysis of the esters 2e and 3e produces respectively the acids 2a and 3a in quantitative yields. All the polymers except 4 are soluble in common solvents, possess high molecular weights (M_w up to ~1.5 × 10⁶ Da), and can be processed into robust engineering forms such as mechanically strong fibers and films.

The polymers are characterized spectroscopically and a few examples of their nuclear magnetic resonance (NMR) spectra are given in Figure 1. The spectrum of 2e in chloroform corresponds well to its expected molecular structure. Isolated amide protons are known to resonate at $\delta < 5.9$, but the resonance of the amide protons of **2e** occurs at $\delta \sim 6.9$, indicating that the amide groups are hydrogen bonded (*11, 16*).

The resonance peaks of **2e** are much broader than those of the poly(phenyl-acetylenes) with no hydrogen bonding moieties (*22, 23*). The line broadening is the result of long spin relaxation time, suggesting that the polymer chains of **2e** are folded and their molecular motions are hence well restricted. In contrast, in acetone, **2e** exhibits sharper spectral peaks (Figure 1B), implying that the polymer takes a relatively extended chain conformation in the polar solvent. Similarly, the polyacid **2a** gives a well-resolved NMR spectrum in a polar solvent of DMSO. The NMR spectra of other amino acid-containing polyacetylenes are also measured in different solvents and a general trend is observed: the polymer solutions in polar solvents [acetone, methanol, tetrahydrofuran (THF), etc.] give better resolved spectra than those in less polar solvents [chloroform, dichloromethane (DCM), benzene, etc.].

Manipulation of Chain Helicity by External Stimuli

All the polymers are cis-rich; for example, the cis contents of **2e** and **2a** estimated from the well resolved spectra shown in panels B and C of Figure 1 are respectively 89.1 and 61.5% (*22, 23*). The cis conformation has two single-bond isomers, i.e. cis-transoid (ct) and -cisoid (cc) (Chart 1A).

In a polar solvent, the solvent molecules may hydrogen bond to the amino acid pendants and the solvated segments may take an extended ct conformation. The pendant groups are, however, too bulky to locate in the same plane and may force the polyacetylene backbone to rotate along the chain axis in order to relieve the steric congestion. In a less polar solvent, the polar amino acid groups may "dislike" the solvent molecules but "like" to self-associate through intrachain hydrogen bonding. Single-bond rotation of the ct conformer generates the cc cousin, which brings the pendant groups to a close vicinity and facilitates the hydrogen bond formation. Similarly, the trans conformation also has two single-bond isomers, trans-cisoid (tc) and -transoid (tt). The chain of the tc conformer is extended but that of the tt conformer is bent because of the steric repulsion of the bulky neighboring pendants locating on the same side of the tt chain.

The formation of the spiral cc and tt chains is entropically unfavorable. The entropic cost may, however, be compensated by the intrachain hydrogen bonds between the amino acid moieties (Chart 1B). When the cooperative force of hydrogen bonds prevails, the α-helix structure is stabilized. In concentrated solutions or solid films, the chain segments may associate with closely located neighbors through interchain hydrogen bonds to form β-sheet structures. As discussed above, the interconversion between the extended and spiral chains is via single-bond rotation (Figure 1A) and the formation of the α-helix and β-sheet structures is through chain interaction (Figure 1B), both of which are sensitive to environmental changes. This offers the opportunity to tune the helicity and to modulate the morphology of the polymers by external stimuli and we thus tried to explore the possibility of "engineering" their folding architectures.

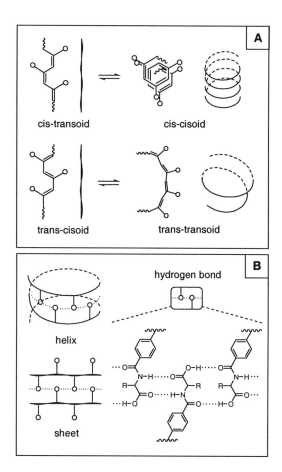

Chart 1. (A) Schematic illustration of cis-transoid (ct), cis-cisoid (cc), trans-cisoid (tc), and trans-transoid (tt) conformations of substituted polyacetylene segments. The ct/tc and cc/tt segments give respectively extended and spiral chains, which are interconvertable via single bond rotation. (B) Graphical representation of the formation of α-helix and β-sheet structures through intra- and interstrand hydrogen bonds of amino acid moieties, respectively.

We first checked how the helicity of the polymer chains changes with solvent. Figure 2A shows the circular dichroism (CD) spectra of **2e** in different solvents (acetone being excluded because it absorbs UV light). In THF, the 1st, 2nd, and 3rd Cotton effects at 374, 315, and 267 nm are (+), (–), and (+) signed, respectively (*24*). In methanol, the CD pattern remains unchanged but the peak intensity weakens, suggesting that part of the helical chains have changed their screw sense. In a less polar solvent of DCM, the CD pattern is completely reversed; that is, the relative population of right- and left-handed helical segments is switched. In chloroform, the preference for one handedness further prevails over the other. In the same solvent, the

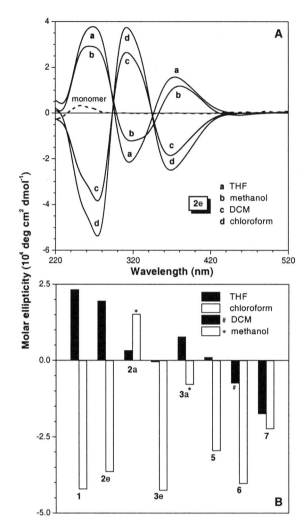

Figure 2. (A) CD spectra of 2e in different solvents (solid lines) and its monomer in chloroform (dotted line). (B) Solvent dependence of the first Cotton effects of the polyacetylene solutions. Concentration: 1.0–2.4 mM. Data of 4 are not shown because its poor solubility precludes accurate measurements of its CD spectra.

monomer displays only a weak CD peak centered at 253 nm. The Cotton effects in the longer wavelength region observed in the polymer solutions thus must be due to the absorption of the polyacetylene chromophores, confirming that the polymer chains are indeed helically rotating.

The solvent dependence of the Cotton effects is not an isolated but a general phenomenon for all the amino acid-containing polyacetylenes. As shown in Figure

2B, for all the polymers, the first Cotton effect in the longest wavelengths changes with solvent. As discussed above, this Cotton effect is associated with the helicity of the backbone segments. Previous studies on the helical polyacetylenes with non-hydrogen bonding chiral pendants found no any solvent effects on their CD spectra (*25, 26*). On the other hand, the induced backbone helicity of poly[4-(carboxy-phenyl)acetylene] $-[HC=C(C_6H_4-p-CO_2H)]_n-$ changed with the type of the chiral amine that hydrogen bonded to the acid groups of the achiral polymer (*6, 24*). The dramatic solvent effects observed in our system thus must have stemmed from the molecular interactions of the solvents with the amino acid pendants, which exerts an asymmetric force field on the polyacetylene backbone. The chirality of the pendant is then transferred to the backbone, causing the segments of the polymer chains to rotate in one preferred screw sense. The rotating direction is determined by the delicate dynamic balance of various interactions, both with the surrounding solvent exterior and within the folded polymer interior, among which, the folding information encoded in the primary structures of the macromolecules by the amino acid moieties plays a pivotal role. This explains why the polymers with different amino acid pendants respond to the solvent variation in different ways; for example, the difference in the responses of **1** and **7** to the same solvent change (THF $\rightarrow$ chloroform) is striking. (Even more drastic difference is observed in the polymers containing sugar moieties such as D-glucose and D-galactose, which will be reported in a separate paper.) Interestingly, however, all the chloroform solutions show negatively signed first Cotton effect, revealing that there is a preferred pattern for the polymer chains to fold in the halogenated solvent.

It is of interest to learn whether the tuning of the chain helicity by solvent is continuous and reversible. Figure 3A shows the CD spectra of **2a** in methanol/chloroform mixtures of varying volume compositions. The methanol solution of **2a** exhibits three alternatively (+), (−), and (+) signed Cotton effects in the visible and UV spectral region. When a small amount of chloroform is admixed with methanol, the CD profile of the polymer solution keeps the same but its peak intensity decreases. Clearly chloroform induces a helical screw sense opposite to that in methanol. The population of the chain segments with this opposite handedness progressively increases with an increase in the volume ratio of chloroform in the mixture and eventually becomes dominant, as revealed by the sign reversal of the CD pattern at high volume fractions of chloroform. In a separate set of experiments where methanol fractions in the mixtures are increased, the CD spectra change in the direction opposite to that discussed above, proving that the helical chirality is reversibly switchable. In other words, the chain segments "remember" their particular conformations in specific environmental conditions, similar to the "memory effect" observed in the achiral polyacid–chiral amine complex systems (*6*). The memory ability of **2a** may arise from the highly specific molecular interaction of the chiral polymer with achiral solvents, a familiar biochemistry topic best exemplified by the unfolding–refolding process of proteins (*1*). Similarly, the helicity of the polyacid **3a** in methanol/chloroform mixtures continuously and reversibly changes with the solvent composition, although the sign inversion of the Cotton effects is not observed in this system (Figure 3B).

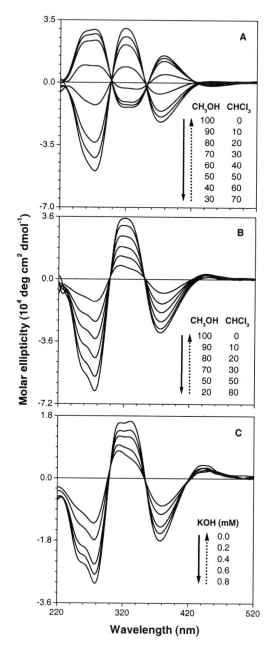

Figure 3. Tuning CD spectra of polyacetylene solutions by (A and B) solvent composition and (C) pH change. Polymer concentration (mM): (A) 1.5–1.7 (2a in methanol/chloroform mixtures), (B) 1.4–1.5 (3a in methanol/chloroform mixtures), (C) 1.2–1.6 (3a in KOH/methanol mixtures).

The biological functions of proteins often change with pH; for example, the enzymatic activity of pepsin peaks around pH 2, while acetylchlolinesterase works best at pH 7 or above (1). This activity dependence is ultimately attributed to the pH-induced change in the folding structures of the proteins. The helical structures of the polyacetylene chains may also change with pH and we thus alkalified the polymer solutions. When a minute quantity of KOH (0.2 mM) is added into a methanol solution of 3a, its CD spectrum is largely intensified (Figure 3C). The Cotton effects continuously increase with an increase in the alkali concentration (or pH value). Addition of acid (HCl) into the KOH solution, however, reclaims the original CD spectrum; that is, the polymer has a "repairing" power and the "denatured" chains refold into the "native" form once the base is neutralized. The chain helicity can also be tuned by additives. For example, addition of 0.3 M of glycine, an achiral amino acid, into a 2a solution (1.7 mM) in a methanol/water (1:1 by volume) mixture doubles the magnitude of the first Cotton effect, although the reversibility is hard to confirm in this system because of the involved technical difficulty.

Morphosynthesis of Natural Hirearchical Structures

In the structural hierarchy of proteins, α helix is a secondary structure, whose change causes further variations in the higher-level structures of the biopolymers. The change in the helicity of the polyacetylene chains should also give rise to variations in their quaternary structures, which is indeed the case, as demonstrated by the atomic force microscopy (AFM) images of the different self-assembling morphologies of 3a formed under different conditions (Figure 4). In the AFM analyses, we allowed the polymer solutions deposited on mica to dry naturally. The formation of the morphologies thus must be very fast, because it needs a split second for a tiny amount (normally 5 µL) of a volatile solvent to evaporate in open air at ambient temperature. In actuality, the high molecular weight polymers should start to "fold" and precipitate on even earlier stage, well before all the solvent molecules evaporate.

When 5 µL of a dilute methanol solution of 3a (12.8 µM) is placed on freshly cleaved mica, twisted cables of helical fibrils is formed, an example of which is shown in Figure 4A. When an extremely dilute solution (2.6 µM) is used, the aggregation of the macromolecules is further suppressed and the resultant morphology looks like long threads or helical chains, each two of which twist over and under each other to form a double helix (inset of Figure 4A). The heights of the helical chains directly measured by AFM and the widths estimated by subtracting the broadening effect of the AFM tips (14, 15) are in the range of ~2.0–2.5 nm, in reasonable agreement with the calculated diameter of 2.3 nm for an extended single polymer chain (20). These morphologies suggest the following pathway for the chain folding: braiding of the single helices gives the double helices, further association of which affords the multistranded cable. In the whole folding process, hydrogen bonds and shape complementarity are believed to serve as the "sticky" adhesives to bundle and link up the polymer chains. Close inspection of the cable structure reveals that the fibrils fold back to form knobs or β bends at two ends of the cable. These hairpin loops are also frequently observed in the morphologies formed by other highly diluted solutions. The β bend is a characteristic way of turning an extended polypeptide chain in a different direction (*1*), as required for compact folding of proteins. Clearly the unnatural polymer is using the same β-turn motif to reverse its chain trajectory.

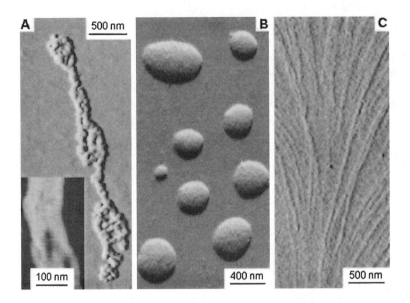

Figure 4. Self-assembling morphologies formed upon natural evaporation of dilute 3a solutions on mica. (A) A multistranded cable from a 12.8 μM methanol solution (inset: duplex braids from a 2.6 μM solution), (B) spherical vesicles from a 9.6 μM chloroform solution with 0.5% methanol, and (C) extended filaments from a 12.8 μM methanol solution containing 25.5 μM KOH.

The morphology obtained from a chloroform solution of similar concentration (9.6 μM) is distinctly different. Instead of forming helical cables, the chloroform solution gives oval eggs or spherical vesicles (Figure 4B). This difference is, however, not totally unexpected. Unlike methanol, chloroform is a poor solvent of amino acids and the polymer chains would self-fold to minimize their exposure to the solvent. When the solvent evaporates, the elementary "foldamers" may cluster together to form compactly packed bigger spheres in an effort to minimize their contact surfaces with air. The morphology does not only change with solvent but also with pH. Addition of KOH to the methanol solution disassembles the helical cables and gives extended filaments (Figure 4C). The complexation of the K^+ ions with the polymer chains partially dissociates the amino acid groups, hindering the hydrogen bond formation. The electrostatic repulsion of the charged chains cannot easily bent but take an extended conformation. It is well known that dilute solutions of polyelectrolytes carrying COO^- ions show unusually high viscosity due to the chain rigidification caused by the imaginary "ion atmosphere" surrounding the polymer chains (27). The extended filament morphology formed by the KOH/3a polyelectrolyte complex clearly demonstrates this effect in a visual fashion. The electrostatic repulsion as well as the steric effect does not allow the charged bulky pendants to stay on the same plane but force them to rotate around the chain axis. This will produce helical chains with short pitches, which may account for the increased Cotton effects of the KOH/3a solutions, cf., Figure 3C (24).

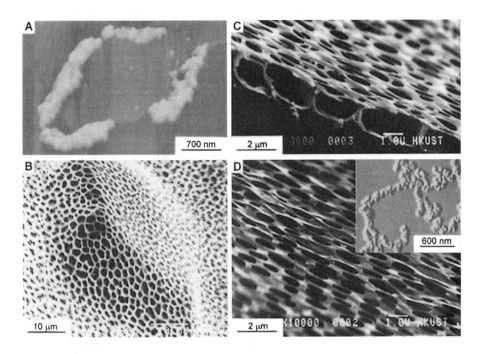

Figure 5. (A) Crescent helices formed by a dilute methanol solution of 3a (12.8 μM). (B) A honeycomb pattern generated from a concentrated THF solution of 3a (28.9 μM), whose 3D nature is clearly revealed by the enlarged side view (C) with the sample tilted 60° from the electron beam axis. (D) Cable (inset) and honeycomb structures formed respectively by dilute (8.2 μM; inset) and concentrated (61.1 μM) THF solutions of 2a.

Nature routinely assembles building blocks into hierarchical structures of aesthetically pleasing patterns with precisely defined biological functions. In sharp contrast, organization of synthetic (macro)molecules into desired high-order structures is still a daunting task for humankind of present day. Development of efficient morphosynthetic processes will bring our construction skill to a new level of structural complexity, namely, from molecular synthesis to morphological fabrication. The encouraging results in tuning the nanoscale morphologies of the polyacetylenes by the external stimuli prompted us to look into the possibility of generating natural organizational structures of higher dimensions. During the AFM study of the morphologies formed by the dilute methanol solutions of 3a, we often observed crescent helix structures. As can be conjured up from the example shown in Figure 5A, there seems to be a few pieces of cables surrounding a drop of solvent; when the solvent evaporates, a fence-like structure consisting of crescent helixes is formed. The closely located helical cables may be interconnected through their "sticky ends" by interstrand hydrogen bonds of the amino acid groups, as implied by the joining corners of the fence structure. The solvent molecules try to separate the polymers apart but the sticky bonds tend to bring them together; the balance of the two antagonistic forces leads to the formation of the fence-like structure.

If the concentration of the polymer solution is increased, more polymer chains will stick together to generate more closed fences or cages and the neighboring cages may merge to form porous structures. When a THF solution of 3a with a relatively high concentration (28.9 μM) is used, a mesoporous film comprising interconnected cages, reminiscent of a honeycomb, is formed (Figure 4B). The thin film spreads over a large area, whose size is only limited by that of the substrate used in the scanning electron microscope (SEM) analysis. The interconnection of the cages is three dimensional in nature, as clearly revealed by the SEM image of the tilted sample. The instant and spontaneous formation of the honeycomb pattern is a regularly observed phenomenon in the evaporation of the solutions of the amphiphilic polyacetylenes with appropriate concentrations, a further example being given in Figure 5D. A concentrated THF solution of 2a (61.1 μM) gives a honeycomb-patterned film spreading over the whole substrate, again with the helical cable fences serving as the structural intermediates.

In organic-inorganic hybrid living systems such as snails, clams, and scallops, cells grow on biomineral surfaces in defined patterns and shapes (2). The mollusk structures can also be generated by the unnatural polymers on nonliving mineral surface, as demonstrated by the snail-shaped morphology formed on depositing a THF solution of 2a on mica (Figure 6A). It is generally accepted that the origin of life is aquatic: the organics produced by the lightning-induced abiotic chemical reactions and/or contained in the meteorite fragments from outer space fell into the shallow lakes and seas that dotted the primitive earth; in those bodies of water life evolved (28). We are intrigued to know what structures the chiral polyacetylenes will adopt when their solutions are poured into water. While 2a solutions are miscible with basic water, the polymer readily precipitates in acidic water. The precipitates have different morphologies, including helical strings (up to millimeters long), mushroom-like fibrillate nets, and amorphous fine particles. Figure 6B shows a partial view of a long fiber of 2a obtained after its methanol solutions have been repeatedly precipitated into slightly acidified water. The fiber is ribbon shaped and helically rotating, resembling a twisted motor nerve winding on the surface of leg muscle (2). The ribbon shape of the helical fiber suggests that the macroscopic structure is formed through side-by-side assembling of helical sheets. Like many helical cables of fibrous proteins (e.g., keratin) (1), the helical fiber of 2a is also left-handed. The microscopic handedness is the consequence of the molecular chirality of the construction components and the shape complementarity of the intermolecular interactions. The ribbon has numerous annuluses running perpendicular to the fiber axis, similar to the outer coat structure of some silk fibers produced by a hornet (*29*). When ether is diffused into a methanol solution of **3a**, left-handed ribbons with an irregular helical periodicity are again formed (Figure 6C). The diffusion process excludes the possibility that the macroscopic helicity is induced by the asymmetric gravity force the polymer chains may have experienced during the precipitation process discussed above.

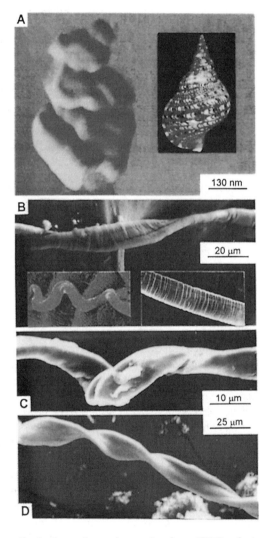

Figure 6. (A) A mollusk shape formed on mica by a THF solution of 2a (18.2 μM). Twisted fibers formed upon (B) precipitation of a methanol solution of 2a (61.1 mM) into acidified water, (C) diffusion of ether into a methanol solution of 3a (3.9 mM), and (D) precipitation of a DMF solution of 4 (100 mM) into an acetone/ether mixture (1:10 by volume). Insets: (A) a snail shell and (B) a twisted motor nerve (left) and a metameric silk fiber (right).

The as-prepared polyacetylenes are completely soluble in their polymerization solvents. All the polymers but **4** remain soluble after purification by one- or two-times precipitation. When the precipitation process further repeats, the polymers gradually become less soluble because more polymer chains have folded into regular structures such as the helical ribbons; **2a**, for example, becomes virtually insoluble in any solvents after three times precipitation. Polymer **4** is an outstanding exception, which cannot even be purified by precipitation due to its great propensity to form the fibrous species. The reason for this is unclear but may be related to the additional cooperative association of the sulfide groups of the methionine pendants. The handedness of the fiber is quite interesting: right- and left-handed segments coexist in one helical ribbon (Figure 6D). This is, however, not a totally strange phenomenon and similar structure exists in nature: right-handed segments of some proteins occasionally end with a residue in left-handed conformation (*1, 2, 28, 30*).

Concluding Remarks

In this study, we have synthesized a group of amphiphilic polyacetylenes, tuned helicity of the polymer chains and their assemblies, and fabricated hierarchical structures through self-assembling processes. Our system has the following noteworthy features:

(1) Different from the aggregates of low molecular weight peptidomimetics and foldamers, the fibrous films of natural structures afforded by our high molecular weight polymers are robust and thus of practical value.

(2) Morphological structures can be formed by block copolymers, which are, however, often synthesized by nontrivial living polymerization reactions under stringent conditions. In contrast, our polymers can be readily prepared by a simple polymerization procedure, which can even be carried out in water in air (*23*).

(3) The helicity of our polymers is much higher than that of their complex cousins; e.g., $[\theta]_{432}$ of an aqueous 3a/Na solution is 14380 deg cm^2 dmol^{-1}, which is >28 times higher than the value reported by Saito *et al.* for their L-leucine–poly[4-(carboxyphenyl)acetylene]/Na complex (**8**) in water (*24*). There is one leucine group in every repeat unit of 3a, whereas in **8**, not every COO$^-$ group can be complexed by leucine due to the well-known steric effect in polymer complexation processes. This structural difference may account for the observed difference in the chain helicity.

(4) It is generally believed that synthetic homopolymers are difficult to form molecular assemblies of defined architecture (*13*). Our polymers can, however, spontaneously assemble into a variety of morphological structures on "all" length scales and spatial dimensions, which is extraordinary, if not unprecedented. The instantaneity of the assembling process is truly amazing: the polymers with molecular weights up to million daltons reproducibly self-organize into specific hierarchical structures in a fraction of a second. Our work clearly demonstrates that synthetic homopolymers with rationally designed molecular structures can be used to fabricate natural morphologies, whose size, shape, and pattern can be readily tuned by simple morphosynthetic processes.

(5) Unlike normal vinyl polymers, our acetylenic polymers consist of alternating double bonds, whose close structural cousins are known to be highly photoconductive (*19*). The helical polyacetylenes thus may find potential applications in artificial nerve systems and phototherapy processes.

Acknowledgements. The work described in this chapter was partially supported by the grants from the Research Grants Council of the Hong Kong Special Administrative Region, China. This project was also benefited from the support of the Technology Resources International Corporation, Georgia, and the Joint Laboratory for Nanostructured materials and Technologies between the Chinese Academy of Sciences and the Hong Kong University of Science & Technology. We thank Drs. Qunhui Sun, Kaitian Xu, and Yuping Dong and Ms. Han Peng of our laboratory, Dr. Jianxiong Li of the Department of Chemical Engineering, Dr. Hong Xue of the Department of Biochemistry, and Dr. Yong Xie and Ms. Jing Zhou of the Department of Biology of our University for their technical assistance and helpful discussions.

Literature Cited

1. Zubay, G. L. *Biochemistry,* 4th ed.; Wm. C. Brown Publishers: Boston, 1998; Chapters 5 and 8.
2. Johnson, G. B *The Living World,* 2nd ed.; McGraw Hill: Boston, 2000; Chapters 4 and 25.
3. Nelson, J. C.; Saven, J. G.; Moore, J. S.; Wolynes, P. G. *Science* **1997**, *277*, 1793.
4. Rowan, A. E.; Nolte, R. J. M. *Angew. Chem. Int. Ed.* **1988**, *37*, 63.
5. Akagi, K.; Piao, G.; Kaneko, S.; Sakamaki, K.; Shirakawa, H.; Kyotani, M. *Science* **1998**, *282*, 1683.
6. Yashima, E.; Maeda, K; Okamoto, Y. *Nature* **1999**, *399*, 449.
7. Zubarev, E. R.; Pralle, M. U.; Li, L. M.; Stupp, S. I. *Science* **1999**, *283*, 523.
8. Jenekhe, S. A.; Chen, X. I. *Science* **1999**, *283*, 372.
9. Green, M. M. *Angew. Chem Int. Ed.* **1999**, *38*, 3139.
10. Whitesides, G. M.; Ismagilov, R. F. *Science* **1999**, *284*, 89.
11. Chung, Y. I.; Christianson, L. A.; Stanger, H. E.; Powell, D. R.; Gellman, S. H. *J. Am. Chem. Soc.* **2000**, *120*, 10555.
12. Cha, J. N.; Stucky, G. D.; Morse, D. E.; Deming, T. J. *Nature* **2000**, *403*, 289.
13. Yu, S. M.; Soto, C. M.; Tirrell, D. A. *J. Am. Chem. Soc.* **2000**, *122*, 6552.
14. Leclère, P.; Calderone, A.; Marsitzky, D.; Francke, V.; Geerts, Y.; Müllen, K.; Brédas, J. L.; Lazzaroni R. *Adv. Mater.* **2000**, *12*, 1042.
15. Lashuel, H. A.; LaBrenz, R.; Woo, L.; Serpell, L. C.; Kelly, J. W. *J. Am. Chem. Soc* **2000**, *122,* 5262.
16. Rivera, J. M.; Craig, S. L.; Martin, T.; Rebek, J.; Jr. *Angew. Chem. Int. Ed.* **2000**, *39*, 2130.
17. Cuccia, L. A.; Lehn, J. -M.; Homo, J. -C.; Schmutz, M. *Angew. Chem. Int. Ed.* **2000**, *39*, 233.

18. Shirakawa, H.; Louis, E. J.; MacDiarmid, A. G.; Chiang, C. K.; Heeger, A. J. *J. Chem. Soc. Chem. Commun.* **1977,** 578.

19. Tang, B. Z.; Chen, H. Z.; Xu, R. S.; Lam, J. W. Y.; Cheuk, K. K. L.; Wong, H. N. C.; Wang, M. *Chem. Mater.* **2000,** *12*, 213.

20. Tang, B. Z.; Kong, X.; Wan, X.; Peng, H.; Lam, W. Y.; Feng, X.; Kwok, H. S. *Macromolecules* **1998,** *31*, 2419.

21. Huang, Y. M.; Lam, J. W. Y.; Cheuk, K. K. L.; Ge, W.; Tang, B. Z. *Macromolecules* **1999,** *32,* 5976.

22. Tang, B. Z.; Kong, X.; Wan, X.; Feng, X. -D. *Macromolecules* **1997,** *30*, 5620.

23. Tang, B. Z.; Poon, W. H.; Leung, S. M.; Leung, W. H.; Peng, H. *Macromolecules* **1997,** *30*, 2209.

24. Saito, M. A.; Maeda, K.; Onouchi, H.; Yashima, E. *Macromolecules* **2000,** *33*, 4616.

25. Ciardelli, F.; Lanzillo, S.; Pieroni, O. *Macromolecules* **1974,** *7*, 174.

26. Moore, J. S.; Gorman, C. B.; Grubbs, R. H. *J. Am. Chem. Soc.* **1991,** *113*, 1704.

27. Elias, H.-G. *Macromolecules: Structure and Properties;* Plenum: New York, 1977; pp 352–354.

28. Brum, G.; McKane, L.; Karp, G *Biology, Exploring Life,* 2nd ed.; Wiley: New York, 1994; Chapter 35.

29. Ishay, J. S.; Kirshboim, S. *Biomacromolecules* **2000,** *1,* 224.

30. Pieroni, O.; Fissi, A.; Pratesi, C.; Temussi, P. A.; Ciardelli, F. *Biopolymers* **1993,** *33*, 1.

Chapter 11

Self-Assembly of Ferrocene-Based Block Copolymers: A Route to Supramolecular Organometallic Materials

K. Nicole Power-Billard, Jason Massey, Rui Resendes, Mitchell A. Winnik[*], and Ian Manners

Department of Chemistry, University of Toronto, 80 St. George Street, Toronto, Ontario M5S 3H6, Canada

Controlled ring-opening polymerization of [1]ferrocenophanes has provided access to poly(ferrocene) block copolymers. These materials undergo phase separation in the solid state and form micellar assemblies that possess organometallic domains in block selective solvents. Possible applications of the materials as precursors to semiconducting or magnetic ceramic nanostructures or as photonic materials are discussed.

Whereas the macromolecular chemistry of carbon is well-developed, the corresponding chemistry of transition metals and main groups elements is much less explored.(1) Nevertheless, previous work has demonstrated that the incorporation of elements from around the periodic table allows access to materials with substantially different properties from those present in organic polymers.(1) Moreover, controlled routes to well-defined inorganic polymers permit access to novel structures which might be used in supramolecular chemistry and the development of "hierarchical" structures with organization on different length scales.

The self-assembly of block copolymers with immiscible segments is known to generate a variety of different morphologies due to phase separation in the solid state and block selective solvation in solution and provides an attractive route to nanostructures.(3-5) Recent examples of block copolymer films and solution micelles containing metal or semiconductor nanoparticles in desired domains provide elegant illustrations of the considerable potential of this approach.(5) In this Chapter we provide an overview of our recent work on novel organometallic nanostructures derived from poly(ferrocene) block copolymers.

Living Anionic ROP of Metallocenophanes

Our group has developed ring-opening polymerization (ROP) routes to high molecular weight poly(ferrocene)s such as poly(ferrocenylsilane) 1 (Figure 1) which display a range of interesting properties.(2-4) Although thermal ROP gives rise to high molecular weight ($M_n > 10^5$) materials there is virtually no chain length control and the molecular weight distributions are broad (typical PDIs = 1.5 - 2.0).

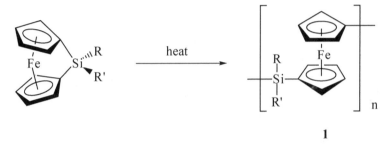

1

Figure 1. Thermal ROP of a [1]silaferrocenophane

We have therefore been very interested in the development of more controlled ROP routes, particularly those which would allow access to multiblock architectures with their associated phase-separated, redox-active domain structures. With this in mind, we reported that silicon-bridged [1]ferrocenophanes undergo living anionic ROP in the presence of initiators such as alkyl and aryllithium reagents.(5,6) This discovery indeed allowed molecular weight control, end group functionalization and access to the first di-, tri- and multiblock copolymers with skeletal transition metal atoms.(6,7) The first "prototypical" diblock materials that we studied were the organometallic-inorganic diblock copolymer poly(ferrocenyldimethylsilane)-*b*-poly(dimethylsiloxane) (PFDMS-*b*-PDMS) and the organic-organometallic diblock copolymer polystyrene-poly(ferrocenyldimethylsilane) (PS-*b*-PFDMS).(6,7) These discoveries in 1994 allowed us to embark on new projects aimed at studies of the self-assembly of these materials to yield organometallic, redox-active nanostructures. We will first focus on our collaborative work on the solution and solid state behavior of these materials.

Self-Assembly of Poly(ferrocenyldimethylsilane)-*block*-Poly(dimethylsiloxane) (PFDMS-*b*-PDMS) in *n*-Hexane

We first discuss the behavior of a PFDMS-*b*-PDMS block copolymer (M_n = 3.7 x 10^4 g mol^{-1}, PDI = 1.10, block ratio 1.0 : 6.0) in *n*-hexane, which is a non-solvent for the PFDMS block.(8) The synthesis of this copolymer is outlined in Figure 2. As expected for diblock copolymers whose components possess a volume ratio in the range of 18 to 40 %, thin films of this block copolymer self-assemble in the bulk state to form hexagonally packed PFDMS cylinders within a PDMS matrix. The solution morphology was investigated by dissolving the block copolymer in *n*-hexane at 80 °C and then visualizing the sample by transmission electron microscopy (TEM). The samples were prepared by atomizing the dilute solution in *n*-hexane using an inert aerosol onto a thin carbon film. The cylinders possess an iron-rich, organometallic core of PFDMS surrounded by an insulating sheath of PDMS. Due to the iron rich PFDMS block, no staining was required to afford contrast. By TEM, regions of close-packed, two-dimensional cylinders were observed in regions in which the polymer solution had been more concentrated. In more dilute regions, isolated cylinders are observed, which possess a variable contour length on the order of 440 nm and a uniform thickness of *ca.* 20 nm (Figure 3). In comparison, visualization by atomic force microscopy (AFM) revealed micellar widths of 68 nm. The difference in the micelle thickness between the two techniques is due to the fact that contrast in TEM is determined by differences in electron density, thus TEM allows only visualization of the iron-rich PFDMS core, whereas AFM permits both the core and the PDMS corona to be observed. Both static and dynamic light scattering (SLS and DLS) experiments confirm the TEM and AFM findings; the mean aggregation number was *ca.* 2000 and a R_G/R_H (radius of gyration to hydrodynamic radius) ratio was 1.2, corresponding to an aspect ratio 5.5, consistent with the cylindrical morphology.

It is possible to tailor the contour lengths of these micelles by varying their preparative technique. First, micelles with average lengths of many microns can be prepared by dialyzing a solution of the block copolymer in THF (a good solvent for both blocks) against *n*-hexane. Second, it is possible to subject the sample to low intensity ultrasonication using a 60-watt sonication bath in order to reduce the micellar contour length. As observed by TEM, the average length was reduced to 77 nm, while SLS and DLS studies indicated that sonication had reduced the aggregation number to 700. Furthermore, R_G/R_H was also reduced to 0.88 with an aspect ratio of 3.1, consistent with the decrease in micelle asymmetry.

As poly(ferrocenylsilane) hompolymers possess interesting hole transport properties with partial oxidation leading to isotropic semiconductivity, it is thus noteworthy that oxidation of the core of these cylindrical structures should lead to their behavior as insulated semiconductor nanowires, in which the PDMS corona acts as an insulator. Upon pyrolysis, such structures might result in magnetic nanostructures with a silicon oxide coating (4,9,10).

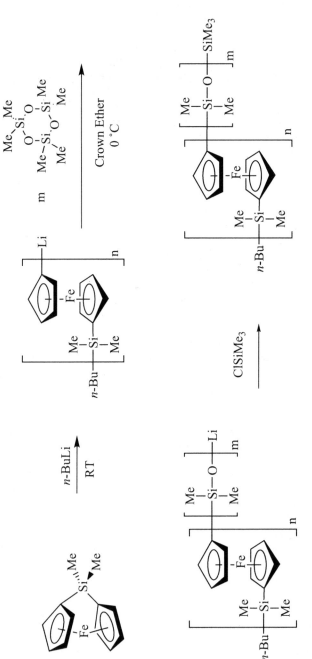

Figure 2. Synthesis of the PFDMS-b-PDMS diblock copolymer

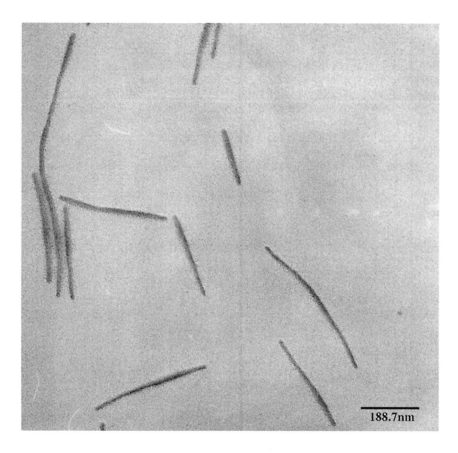

188.7nm

Figure 3. Transmission electron micrograph of cylindrical micelles of PFDMS-b-PDMS (M_n = 37 000 g mol^{-1}, PDI = 1.10) aerosol-sprayed from n-hexane

Self-Assembly of Polystyrene – *block* – Polyferrocene Diblock Copolymers

Another prototypical block copolymer synthesized by our group was the organic-organometallic block copolymer polystyrene-*b*-poly(ferrocenyldimethylsilane), PS-*b*-PFDMS. Both the solution and solid state morphologies of this system have been investigated.(12) A typical PS-*b*-PFDMS diblock copolymer was synthesized (Figure 4) via anionic polymerization with a block ratio of 9.0 : 1.0 (M_n = 1.3 x 10^4 g mol^{-1}, PDI = 1.16). In order to be observed by AFM, a dilute solution of the polymer in acetone (a selective solvent for the low molecular weight polystyrene block) was aerosol-sprayed onto a freshly cleaved mica surface. Spherical micelles (average diameter = 20 nm) were observed in which the PS block formed the corona with the PFDMS as the core-forming block.

The solid state morphology of PS-*b*-PFDMS has also been explored. As a consequence of the crystallinity of the PFDMS block, our studies show that samples required annealing above the melting point of the PFDMS block (120 – 145 °C) to induce ordering. In contrast, thin films of polystyrene-*b*-poly(ferrocenylmethylphenylsilane) (PS-*b*-PFMPS), which contains an amorphous and atactic PFMPS block, exhibit a regular microphase segregated structure without requiring annealing at high temperatures.

Transition-Metal Catalyzed ROP: A Convenient Alternative to Anionic Polymerization

As illustrated in the previous sections, sequential living anionic polymerization provides the most common route towards the synthesis of block copolymers. However, due to synthetic challenges of anionic polymerization (*e.g.* stringent purity requirements for all solvents, monomers, etc.), it is of considerable interest to pursue less experimentally difficult and less time consuming routes to block copolymers. In 1998 we reported the synthesis of multiple poly(ferrocene) block copolymer architectures via a facile transition-metal catalyzed ROP approach.(13) Using this methodology, poly(ferrocene) blocks are "grown" from reactive Si-H functionalities by the addition of the [1]ferrocenophane monomer to a commercially available Si-H end functionalized polymer in the presence of a Pt(0) (Karstedt's) catalyst in order to obtain either a di- or triblock copolymer. We have recently shown that by using Pt(0)-catalyzed ROP of a strained [1]silaferrocenophane in the presence of Si-H terminated poly(ethylene oxide), a water-soluble amphiphilic poly(ethylene oxide)-*b*-poly(ferrocenyldimethylsilane) diblock copolymer is accessible.(14) The polymer was found to self-assemble in water to form spherical micelles with an average diameter of *ca.* 150 nm (Figure 5). We will herein focus on using this same methodology to create triblock poly(ferrocenyldimethylsilane)-*b*-poly(dimethylsiloxane)-*b*-poly(ferrocenyldimethylsilane) (PFDMS-*b*-PDMS-

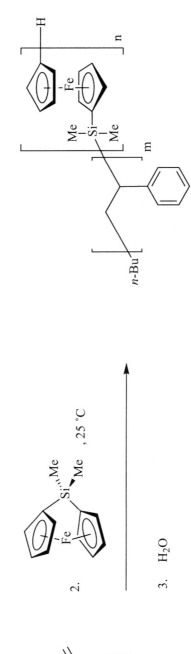

Figure 4. Synthesis of the PS-b-PFDMS diblock copolymer

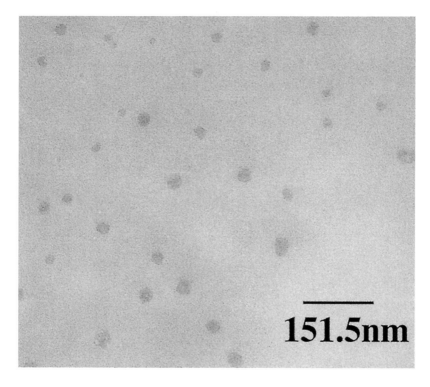

Figure 5. Transmission electron micrograph of spherical micelles of PEO-b-PFDMS aerosol-sprayed from water

b-PFDMS) triblock copolymers which self-assemble to afford a novel flower-like morphology in *n*-hexane solution.(15)

Synthesis and Self-Assembly of PFDMS-*b*-PDMS-*b*-PFDMS Triblock Copolymers

The triblock structure of the polymer was confirmed by synthesizing a low molecular weight PFDMS-*b*-PDMS-*b*-PFDMS block copolymer as a model (Figure 6). The polymer was isolated as an orange powdery material ($M_n = 8.14 \times 10^3$ g mol^{-1}, PDI = 1.45) while ^{1}H and ^{29}Si NMR confirmed the proposed structure. The polymer possessed a PFDMS : PDMS : PFDMS block ratio of approximately 1 : 7 : 1, as indicated by ^{1}H NMR integration. A higher molecular weight triblock copolymer with a long PDMS block was then prepared as an orange gum ($M_n = 2.88 \times 10^4$ g mol^{-1}, PDI = 1.43). ^{1}H NMR integration for this material indicated an approximate PFDMS : PDMS : PFDMS block ratio of 1 : 13 : 1. Consistent with phase separation in the solid state, DSC analysis revealed thermal transitions close to those of the constituent homopolymers ($T_g = 26$ °C : PFDMS blocks (c.f 33 °C for PFDMS homopolymer), $T_g = -123$ °C, $T_{cryst} = -103$ °C, $T_m = -44$ °C : PDMS blocks).

The solution aggregation of the high molecular weight triblock copolymer was studied in *n*-hexane, a good solvent for PDMS and a precipitant for PFDMS. A micellar solution was then prepared by first dissolving the polymer in THF, a good solvent for each block, slowly adding *n*-hexane until the onset of micelle formation, and subsequently dialyzing against *n*-hexane. After solvent evaporation, TEM studies of this micellar solution revealed three main, co-existent morphologies. Specifically, spherical micelles, cylindrical micelles and also novel, flower-like structures were observed (Figure 7).

Fractionation was performed in order to investigate the influence of the composition variation on the observed morphology present in samples of the triblock copolymer. Centrifugation (1.5×10^4 rpm. for 20 min.) of the dialyzed *n*-hexane solution of the polymer afforded a two phase system with a very light, amber supernatant solution and a yellow precipitate. TEM analysis of the supernatant after solvent evaporation revealed that it was composed almost entirely of spherical micelles. ^{1}H NMR integration of the sample gave an approximate PFDMS : PDMS : PFDMS block ratio of 1 : 60 : 1. This is indicative of a reduced PFDMS content compared to the unfractionated sample. GPC analysis revealed this fraction to be of lower molecular weight ($M_n = 2.52 \times 10^4$, PDI = 1.50). In contrast, TEM analysis of an *n*-hexane solution of the above mentioned yellow precipitate, after solvent evaporation, revealed that it contained predominantly flower-like and short cylindrical micelles. ^{1}H NMR integration of this sample revealed a PFDMS : PDMS : PFDMS block ratio of approximately 1 : 6 : 1, which is consistent with a significantly higher PFDMS content compared to the unfractionated sample. GPC analysis of this fraction revealed a higher molecular weight ($M_n = 3.97 \times 10^4$ g mol^{-1}, PDI = 1.37). These

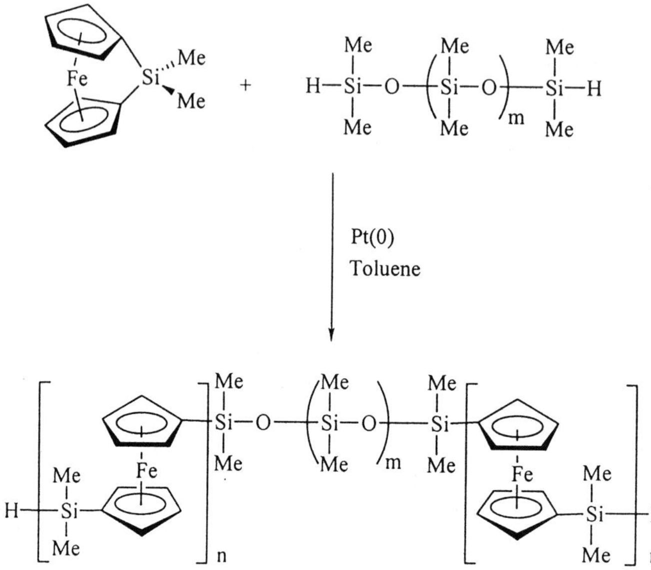

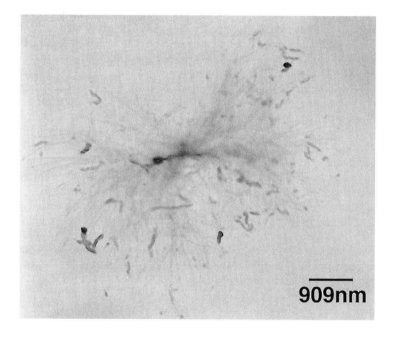

Figure 6. Synthesis of the PFDMS-b-PDMS-b-PFDMS triblock copolymer

Figure 7. Observed morphology by transmission electron microscopy of the PFDMS-b-PDMS-b-PFDMS triblock copolymer in n-hexane

observations suggest that the variation in the relative block lengths present in the sample is critical in determining the observed micellar morphologies in this organometallic-inorganic triblock system.

Future Directions and For Poly(ferrocenes) and Poly(ferrocene) Block Copolymers: Applications as Nanostructured Materials

It is intriguing to gaze into the crystal ball and to speculate with respect to some of the areas in which poly(ferrocene)s and poly(ferrocene) block copolymers are likely to have future impact. A variety of ideas under investigation now follow and in several cases, preliminary results have already been obtained.

One area of considerable potential involves the formation of magnetic nanostructures. Poly(ferrocene)s are known to function as preceramic polymers and yield interesting ferromagnetic Fe-containing ceramic composites at 500 - 1000°C.(4) The use of such involatile but processable polymeric precursors to ceramics is a potentially attractive way of circumventing the difficulty of processing ceramic materials into desired shapes. We have been intrigued by the possibility of making magnetic nanostructures from nanodomains of poly(ferrocenylsilane)s and the first successful results were reported in early 1998. In collaboration with the group of Prof. G. Ozin at Toronto we have explored the pyrolysis of poly(ferrocenylsilane)s within the 30 – 40 Å channels of mesoporous silica, MCM-41.(4,9,10) The monomer can be sublimed into the hexagonal mesopores at room temperature and ROP can be induced by heating the resulting material at 200°C. Heating to 900°C generates black, nanostructured magnetic ceramic products which can be visualized in the channels by TEM. The materials are superparamagnetic as the channels restrict the growth of the iron clusters so that they are too small to permit stable magnetization. Magnetic measurements, X-ray diffraction, and TEM confirm that the composites display superparamagnetic behavior consistent with the generation of very small iron particles (c.a. 20 – 40 ± 5 Å size) in a SiC/C matrix.(10)

Layer-by-layer assembly of aqueous organic polyelectrolytes has been widely explored as a method for the fabrication of electrostatic superlattices with a variety of applications including thin film devices and lithography. The first water soluble poly(ferrocenylsilane)s have recently been prepared by our group and with cationic materials and organic polyanions layer by layer self-assembly on surfaces such as Au and Si has been successfully achieved.(16,17) This work on "organometallic superlattices" has been performed in collaboration with the group of Prof. G. Ozin and the resulting structures are of interest for a variety of applications.

As outlined earlier in the chapter, one idea involves the generation and study of spherical, cylindrical and related self-assembled micellar structures (see above) as precursors to semiconducting nanostructures by oxidation or magnetic nanostructures by pyrolysis (Figure 8). Films of poly(ferrocene) block copolymers, which undergo phase separation to generate spherical, cylindrical, and lamellar domains (see above), allows access to highly ordered, periodic redox-active organometallic nanostructures which could be selectively oxidized (chemically or electrochemically *e.g.* by using scanning probe microscope tips) to give materials with interesting properties. For example, the controlled orientation of nanoscale block copolymer domains might allow anisotropic charge transport or the generation of magnetic nanostructures via oxidation or pyrolysis.

The selective ablation of blocks using reactive ion etching (*e.g.* O_2, Ar, H_2 etc) allows the generation of periodic and aperiodic structures of dimensions < 100 nm, less than those readily available using conventional lithographic methodologies. Such techniques have been successfully used, for example, by Chaikin, Register *et al.*, Thomas *et al.* and by Möller and coworkers where inorganic components in one block allow the etching contrast.(18,19,20,21) Poly(ferrocene)s such as poly(ferrocenylsilane)s possess inorganic Fe and Si components which, in the case of easily ablated organic co-blocks, allows high selective etching contrasts (*e.g.* via the formation ablation-resistant oxides using oxygen reactive ion etching). This approach offers opportunities for nanolithography in which periodic and aperiodic structures of dimensions < 100 nm are generated (Figure 8). In collaboration with J. Spatz and M. Möller we have found that the organic components of PS-*b*-PFDMS block copolymer films can be ablated using an H_2-plasma to leave ceramic, Fe-containing residues.(22,23,24) Poly(ferrocenylphosphine)s, which we have also prepared by anionic ROP, possess coordinating phosphorus sites in the main chain and additional metallization should be possible. This provides further interest from this type of patterning perspective.(25)

An interesting extension of these ideas is to use ordered poly(ferrocene) block copolymer films (e.g. spherical and lamellar structures) as photonic crystals in which periodic variations in refractive index are present. These have many potential applications in an array of devices such as waveguides and reflectors. Large dielectric contrast between the blocks is a potential advantage of using poly(ferrocene) materials as the refractive indices of the latter are high (n = ca. 1.7).(23)

Acknowledgment   We thank coworker Karen Temple whose research is reviewed in this article. Much of the work was achieved in collaboration with Prof. Geoffrey A. Ozin (magnetic nanostructures) of our Department and Prof. Mark Foster (Department of Polymer Science, Akron; thin film morphology of block copolymers).

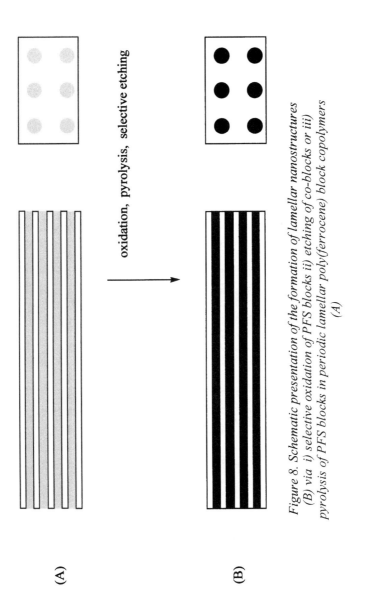

Figure 8. Schematic presentation of the formation of lamellar nanostructures (B) via i) selective oxidation of PFS blocks ii) etching of co-blocks or iii) pyrolysis of PFS blocks in periodic lamellar poly(ferrocene) block copolymers (A)

162

References

1. Manners, I. *Angew. Chem. Int. Ed. Engl.*, **1996**, 35, 1602.
2. Foucher, D. A.; Tang, B. Z.; Manners, I. *J. Am. Chem. Soc.*, **1992**, *114*, 6246.
3. Manners, I. *Can. J. Chem.* **1998**, *76*, 731.
4. Manners, I. *Chem. Commun.* **1999**, 857.
5. Rulkens, R.; Lough, A. J.; Manners, I. *J. Am. Chem. Soc.*, **1994**, *116*, 797.
6. Rulkens, R.; Ni, Y.; Manners, I. *J. Am. Chem. Soc.* **1994**, *116*, 12121.
7. Ni, Y.; Rulkens, R.; Manners, I. *J. Am. Chem. Soc.* **1996**, *118*, 4102.
8. a) Massey, J.; Power, K. N.; Manners, I.; Winnik, M. A. *J. Am. Chem. Soc.* **1998**, *120*, 9533. b) Massey, J.; Temple, K.; Cao, L.; Raez, J.; Winnik, M. A.; Manners, I. *J. Am. Chem. Soc.*, **2000** submitted.
9. MacLachlan, M. J.; Aroca, P.; Coombs, N.; Manners, I.; Ozin, G.A. *Adv. Mater.*, **1998**, *10*, 144.
10. MacLachlan, M. J.; Ginzburg, M.; Coombs, N.; Raju, N. P.; Greeden, J. E.; Ozin, G. A.; Manners, I. *J. Am. Chem. Soc.* **2000**, *122*, 3878.
11. MacLachlan, M. J.; Ginzburg, M.; Coombs, N.; Coyle, T. W.; Raju, N. P.; Greeden, J. E.; Ozin, G. A.; Manners, I. *Science* **2000**, *287*, 1460.
12. Sheller, N.; Li, W.; Foster, M. D.; Balaishis, D.; Manners, I.; Annis, B.; Lin, J-S. *Polymer* **2000**, *41/42*, 719.
13. Goméz-Elipe, P.; Resendes, R.; Macdonald, P. M.; Manners, I. *J. Am. Chem. Soc.* **1998**, *120*, 8348.
14. Resendes, R.; Massey, J.; Dorn, H.; Winnik, M. A.; Manners, I. *Macromolecules* **2000**, *33*, 8.
15. Resendes, R.; Massey, J.; Dorn, H.; Power, K. N.; Winnik, M. A.; Manners, I. *Angew. Chem. Int. Ed. Engl.* **1999**, *38*, 2570.
16. Power-Billard, K. N.; Manners, I. *Macromolecules,* **2000**, *33*, 26.
17. Ginzburg, M.; Galloro, J.; Jäkle, F.; Power-Billard, K. N.; Manners, I.; Ozin, G. *Langmuir,* **2000**, in press.
18. Spatz, J. P.; Mößmer, S.; Möller, M.; Herzog, T.; Plettl, A.; Ziemann, P. J. *J. Lumines.* **1998**, *76/77*, 168.
19. Spatz, J. P.; Herzog, T. ; Mößmer, S.; Ziemann, P.; Möller, M. *Adv. Mater.* **1999**, 11, 149.
20. Park, M.; Harrison, C.; Chaikin, P. M.; Register, R. A.; Adamson, D. A. *Science* **1997**, *276*, 1401.
21. Avgeropoulos, A.; Chan, V. Z-H; Lee, V. Y.; Ngo, D.; Miller, R. D.; Hadjichristidis, N.; Thomas, E. L. *Chem. Mat.* **1998**, *10*, 2109.
22. Massey, J.; Spatz, J.; Möller, M.; Winnik, M. A.; Manners, I., unpublished results
23. Manners, I. *Pure. App. Chem.* **1999**, *71*, 1471.
24. For related work, see: a) Massey, J.; Winnik, M. A.; Manners, I.; Chan, V. Z-H.; Spatz, J. P.; Ostermann, M.; Enchelmaier, R.; Möller, M. submitted. b) Lammertink, R. G. H.; Hempenius, M. A.; van den Enk, J. E.; Chan, V. Z.-H.; Thomas, E. L.; Vancso, G. J. *Adv. Mater.* **2000**, *12*, 98.
25. Cao, L.; Winnik, M. A.; Manners, I. *J. Inorg. Organomet. Polym.* **1998**, *8*, 215.

Chapter 12

Toward Functional Architectures via Terpyridine-Based Metallo-Supramolecular Initiators

U. S. Schubert[1,2,3,*], G. Hochwimmer[1], and M. Heller[1,2]

[1]Lehrstuhl für Makromolekulare Stoffe, Technische Universität München, Lichtenbergstrasse 4, D-85747 Garching, Germany
[2]Center for NanoScience, Ludwig-Maximilians-Universität München, Geschwister-Scholl-Platz 1, D-80593 München, Germany
[3]Laboratory of Macromolecular and Organic Chemistry, Eindhoven University of Technology, P.O. Box 513, NL-5600 MB Eindhoven, The Netherlands

2,2':6',2"-Terpyridine metal complexes functionalized with bromomethyl groups in 5,5"- and 4'-position were used as supramolecular initiators for the living polymerization of 2-oxazoline monomers. Polymers with narrow molecular weight distribution were obtained and the molecular weight could be predetermined by the monomer to initiator ratio. Defined functional endgroups could be introduced via termination of the living chains. The central metal ions such as iron(II) or cobalt(II) could be removed under basic conditions resulting in uncomplexed polymers with free terpyridine segments. After addition of a metal salt almost complete recomplexation to the supramolecular systems occurs. The polymers were characterized using GPC, UV/VIS spectroscopy, MALDI-TOF-MS and NMR.

The controlled fabrication of ordered architectures on a nano- to mesoscopic length scale is one major topic in current polymer chemistry, supramolecular chemistry and material science [1]. Such systems, e.g. on surfaces, in thin films or in bulk materials, could provide new thermal, mechanical, photochemical, electrochemical or magnetic properties and find potential applications in nanotechnology, coatings, intelligent glues or functional surfaces. Systems containing transition metal ions are of particular interest due to the special properties of the metal complexes. All of these attempts require a very precise control of the structures at very different length scales from molecular size to micrometers. At present, several quite different approaches to such materials are employed: A): In polymer chemistry phase separation of block copolymers or the formation of defined micelles is being used [2]. B): Material scientists increase the precision for the mirostructuring of surfaces or films [3]. C): In supramolecular chemistry the self-assembly of suitable organic or inorganic molecules via non-covalent interactions is utilized.

The combination of supramolecular building blocks and polymer chemistry utilizing metallo-supramolecular initiators for living or controlled polymerization reactions seems to be one of the most promising approaches. In the present paper, we describe the extension of our previous efforts concerning the living polymerization of 2-oxazolines utilizing metallo-supramolecular initiators based on different functionalized terpyridines (**Figure 1**). Transition metal complexes of terpyridines play an important role in some promising fields of modern chemistry and physics, due to their wide range of electrochemical [4], photochemical [5] and bio-catalytic [6] properties. In addition, they offer the possibility of a combination with bio-molecules [7].

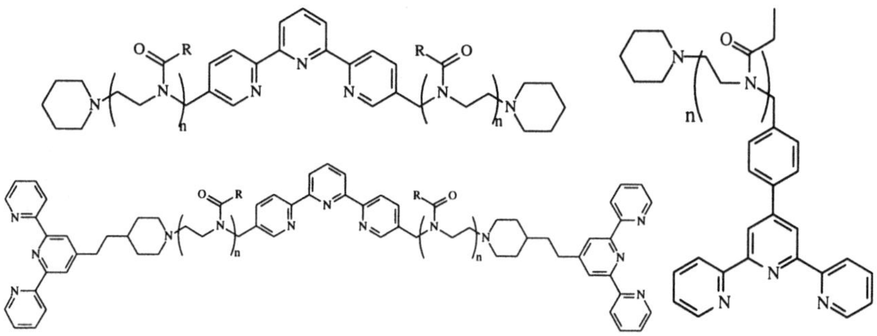

Figure 1. Schematic presentation of uncomplexed terpyridine poly(oxazoline)s with different end groups functionalized in the 5,5''-position (left) and the 4'-position (right).

Experimental Section

Preparation of functionalized terpyridines:

5,5''-Bis(bromomethyl)-2,2':6',2''-terpyridine (see [19] for the synthesis): Yield: 24%. M.p. 195-196°C (188°C [8]). ^{1}H NMR (CDCl$_3$, 300 MHz): δ (ppm) 4.56 (s, 4 H), 7.92 (dd, 2 H, J = 8.39, 2.29 Hz), 7.97 (t, 2 H, J = 8.01 Hz), 8.47 (d, 2 H, J = 8.02 Hz), 8.61 (d, 2 H, J = 8.01 Hz), 8.72 (d, 2 H, J = 2.29 Hz). ^{13}C NMR (CDCl$_3$, 75 MHz): δ (ppm) 29.54, 121.24, 121.55, 133.87, 137.79, 138.08, 149.06, 154.61, 155.78. EI MS, m/z 419 (20%, M + 1). Anal. Calc. for C$_{17}$H$_{13}$N$_3$Br$_2$: C, 48.72; H, 3.13; N, 10.03. Found: C, 48.63; H, 2.68; N, 10.05.

4'-(4-Bromomethylphenyl)-2,2':6',2''-terpyridine (see also [9], [10]): A mixture of 5.0 g (15.5 mmol) 4'-(4-bromomethylphenyl)-2,2':6,2''-terpyridine, 3.3 g (18.5 mmol) N-bromosuccinimide (NBS)(recrystallized from water) and 200 mg, 1.2 mmol α,α'-azoisobutyronitrile (AIBN) were dissolved in dry tetrachloromethane (50 ml) under nitrogen. The mixture was refluxed for 2 h and then filtered while still hot. After removal of the CCl$_4$, the precipitate was recrystalized from dichloromethane to yield 3.8 g of a white crystalline solid (67%, Lit. 72% [10]). M.p. 162°C. ^{1}H NMR (CDCl$_3$, 300 MHz): δ (ppm) 4.75 (s, 2 H), 7.37 (dd, 2 H, J = 7.2, 5.0 Hz), 7.54 (d, 2 H, J = 8.0 Hz), 7.90 (m, 4 H), 8.68 (d, 2 H, J = 8.0 Hz), 8.74 (m, 4 H). ^{13}C-NMR (CDCl$_3$, 75 MHz): δ (ppm) 33.30, 119.32, 121.84, 124.30, 128.03, 128.16, 130.02, 137.45, 138.92, 139.05, 149.34, 156.22, 156.36. Anal. calc. for C$_{22}$H$_{16}$BrN$_3$: C, 65.68; H, 4.01; N, 10.45. Found: C, 65.62; H, 3.71; N, 10.36. UV/VIS (CH$_3$CN): λ_{max}/nm (ε/(L × mol^{-1} × cm^{-1})) = 252 (15746), 278 (19571).

Preparation of the metallo-supramolecular initiators:

Bis(5,5''-bis(bromomethyl)-2,2':6',2''-terpyridine) iron(II) hexafluorophosphate (see also [19]): To a solution of 246.7 mg (0.589 mmol) 5,5''-bis(bromomethyl)-2,2':6',2''-terpyridine in methanol (80 mL) was added 81.9 mg (0.295 mmol) FeSO$_4$ × 7 H$_2$O suspended in methanol (15 mL). The reaction mixture turned immediately to red color. After stirring at room temperature under nitrogen for 8 h, excess NH$_4$PF$_6$ (2.5 g, 15 mmol) was added (solution in methanol) leading to a precipitation of a red solid and the mixture was further stirred for 5 min. After 24 h the solid was filtered off and washed with cold methanol (5 mL), water (20 mL) and diethylether (10 mL) followed by a recrystallization from acetonitrile/diethylether yielding 282 mg (80%) of the desired product as a red cristalline solid. M.p. 225-226°C. ^{1}H NMR (CD$_3$CN, 300 MHz): δ (ppm) 4.23 (s, 8 H), 7.00 (s, 4 H), 7.97 (d, 4 H, J = 8.40 Hz), 8.48 (d, 4 H, J = 8.39 Hz), 8.75 (t, 2 H, J = 8.01 Hz), 8.96 (d, 4 H, J = 8.01 Hz). ^{13}C NMR (CD$_3$CN, 75 MHz): δ (ppm) 29.34, 125.15, 125.77, 139.92, 141.08, 154.18, 158.95, 161.50, 178.53. Anal. Calc. for C$_{34}$H$_{26}$Br$_4$N$_6$FeF$_{12}$P$_2$: C, 34.49; H, 2.21; N, 7.10. Found:

C, 34.77; H, 2.52; N, 7.06. UV/VIS (CH$_3$CN): λ_{max}/nm (ε/(L × mol^{-1} × cm^{-1})) = 561 (6400), 328 (62000), 280 (48900).

Cobalt(II)-bis-4'-(4-bromomethylphenyl)-2,2':6',2''-terpyridine-hexafluorophosphate: To a solution of 4'-(4-bromomethylphenyl)-2,2':6',2''-terpyridine (500 mg, 1.24 mmol) in 100 ml methanol (abs.) a solution of cobalt(II)-acetate (154 mg, 0.62 mmol) in 30 ml methanol (abs.) was added. The solution was then stirred under nitrogen for 2 h at room temperature. To the reaction mixture ammonium hexafluorophosphate (3.0 g, 18 mmol, 29 eq.) dissolved in 10 ml methanol was added. An orange solid precipitated. The solid was filtered off, washed with methanol (3 × 30 ml), H$_2$O (3 × 30 ml) and diethylether (3 × 30 ml). The product was dissolved in acetone and precipitated once again in diethylether. After filtration 450 mg of cobalt(II)-*bis*-4'-(4-bromomethylphenyl)-2,2':6',2''-terpyridine hexafluorophosphate was obtained (63%). ^{1}H NMR (CD$_3$CN, 300 MHz): δ (ppm) 4.79 (s, 4 H), 7.08 (t, 4 H, J = 6.68 Hz), 7.19 (d, 4 H, J = 4.19 Hz), 7.86 (d, 4 H, J = 8.58 Hz), 7.90 (t, 4 H, J = 8.58, 7.83 Hz), 8.29 (d, 4 H, J = 8.2 Hz), 8.60 (d, 4 H, J = 7.82 Hz), 9.17 (s, 4 H). ^{13}C-NMR (CD$_3$CN, 75 MHz): δ (ppm) 32.53, 121.25, 123.52, 127.02, 127.61, 128.12 128.37, 130.12, 136.41, 138.45, 152.78, 157.70, 160.05. Anal. calc. for C$_{44}$H$_{32}$Br$_2$CoF$_{12}$N$_6$P$_2$ × H$_2$O: C, 45.07; H, 2.90; N, 7.17. Found: C, 44.57; H, 2.80; N, 7.12. UV/VIS (CH$_3$CN): λ_{max}/nm (ε/(L × mol^{-1} × cm^{-1})) = 285 (56417).

Iron(II)-bis-4'-(4-bromomethylphenyl)-2,2':6',2''-terpyridine-hexafluorphosphate: The same preparation method as described above was used utilizing iron(II)-sulfate heptahydrate as metal salt. Yield: 86%. ^{1}H NMR (CD$_3$CN, 300 MHz): δ (ppm) 4.80 (s, 4 H), 7.08 (d, 4 H, J = 6.87 Hz), 7.17 (d, 4 H, J = 4.77 Hz), 7.86 (d, 4 H, J = 8.39 Hz), 7.87 (t, 4 H, J = 8.39, 7.63 Hz), 8.29 (d, 4 H, J = 7.82 Hz), 8.60 (d, 4 H, J = 8.01 Hz), 9.17 (s, 4 H). ^{13}C-NMR (CD$_3$CN, 75 MHz): δ (ppm) 32.53, 121.26, 123.53, 126.99, 127.61, 128.12, 128.37, 130.12, 136.40, 138.45, 152.78, 157.70, 159.98. Anal. calc. for C$_{44}$H$_{32}$Br$_2$FeF$_{12}$N$_6$P$_2$: C, 45.90; H, 2.78; N, 7.30. Found: C, 46,42; H, 2.50; N, 7.62. UV/VIS (CH$_3$CN): λ_{max}/nm (ε/(L × mol^{-1} × cm^{-1})) = 285 (46308), 568 (15086).

Polymerization:
The metallo-supramolecular initiator was dissolved in dry acetonitrile and the 2-ethyl-2-oxazoline monomer was added. The reaction mixture was stirred at 80°C for 24 h. After consumption of the monomer piperidine or other termination agents were added and the mixture was stirred for another 5 h. The mixture was allowed to cool to room temperature and the solvent was evaporated *in vacuo*, the polymer was dissolved in CH$_2$Cl$_2$ and precipitated in cold diethylether. After filtration under nitrogen, precipitation was repeated twice. Then the polymer was filtrated off under nitrogen and dried *in vacuo*.

Extraction of the central metal ion:
The polymer (100 mg) was dissolved in 50 mL CH_3CN/H_2O (1:1) and K_2CO_3 (5 g) was added. The mixture was refluxed for 3 h and the layers were separated. The CH_3CN phase was evaporated *in vacuo* and the remaining polymer was dissolved in CH_2Cl_2 and precipitated in diethylether.

Characterization:
Gel permeation chromatography (GPC) analysis was performed on a Waters Liquid Chromatograph system using Shodex GPC K-802S columns, Waters Differential Refractometer 410 and Waters UV Absorption Detector 486 with chloroform as eluent. Calibration was conducted with polystyrene standards. UV/VIS measurements were recorded using a Varian Cary 3 UV/VIS spectrometer. MALDI-TOF-mass spectra were measured with Bruker Biflex 3 mass spectrometer (matrix dithranol, solvent acetone).

Results and Discussion

A promising approach for the combination of metallo-supramolecular building blocks and polymer chemistry is the application of supramolecular units as initiators for living or controlled polymerization reactions (for recent results utilizing controlled radical polymerizations, see e.g. [11] and for living anionic polymerization, see e.g. [12]). In particular, the living cationic polymerization of 2-oxazolines seems to be perfectly suited for this purpose, due to a broad range of different polymer structures that are obtainable by using different monomers [13]. We recently described a new approach towards bipyridine containing polymers utilizing metallo-supramolecular initiators for the living cationic polymerization of 2-oxazolines based on *bis*functionalized 6,6'-dimethyl-2,2'-bipyridine [14] and monofunctionalized 5,5'-dimethyl-2,2'-bipyridine [15] metal complexes (see also results by *Fraser et al.* [16]). The living nature of the polymerization could be demonstrated by the linear relationship between the [monomer]/[initiator] ratio and the average molecular weight, as well as the narrow molecular weight distribution of the polymers. Besides homo polymers, block copolymers with amphiphilic properties and supramolecular segments based on different 2-oxazoline monomers with hydrophilic or hydrophobic characteristics could be obtained [17].

Terpyridines are *N*-heterocyclic ligands with even stronger binding constants for octahedral metal ions [4-6, 18]. We have recently studied *bis*-functional 5,5''-dimethyl-2,2':6',2''-terpyridine metal complexes as initiators for supramolecular 2-oxazoline-polymers [19] (**Figure 2**). Besides the demonstration of the living character of the polymerization, end group functionalization and block copolymerization, we could show that the central metal ion could be removed under basic conditions. First studies combining UV/VIS spectroscopy and cyclovoltammetric measurements revealed that it is possible to perform a partially reversible electrochemical "switch" between the star-like complexed architecture and the uncomplexed polymer species [20].

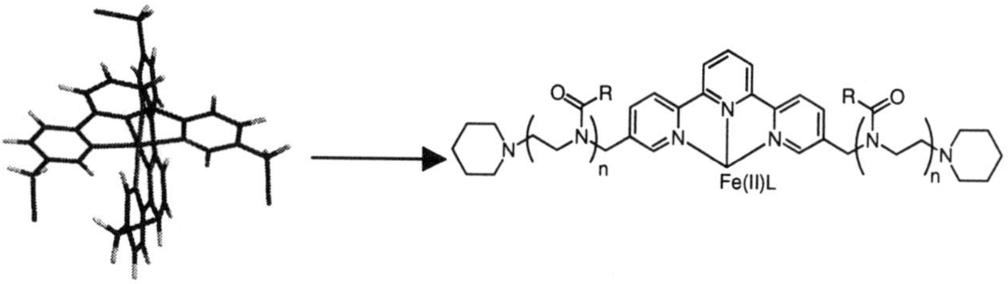

Figure 2: Polymerization procedure towards poly(oxazoline)s containing 5,5"-bisfunctionalized 2,2':6',2"-terpyridines utilizing 2-oxazoline monomers and termination with piperidine (L refers to another terpyridine ligand).

In order to extend this approach and to incorporate monofunctionalized terpyridine ligands, we decided to use the well-known 4'-functionalized terpyridine ligands as building blocks for metallo-supramolecular initiators (for different applications of this building unit see, e.g. [9]). For that purpose, we developed a high yield functionalization of the commercially available 4'-(4-methylphenyl)-2,2':6',2"-terpyridine using radical bromination procedures (**Figure 3**) (for other functionalization reactions, see [9, 10, 21]). The functionalized ligands can easily be complexed with a wide range of octahedral complexing metal ions [22].

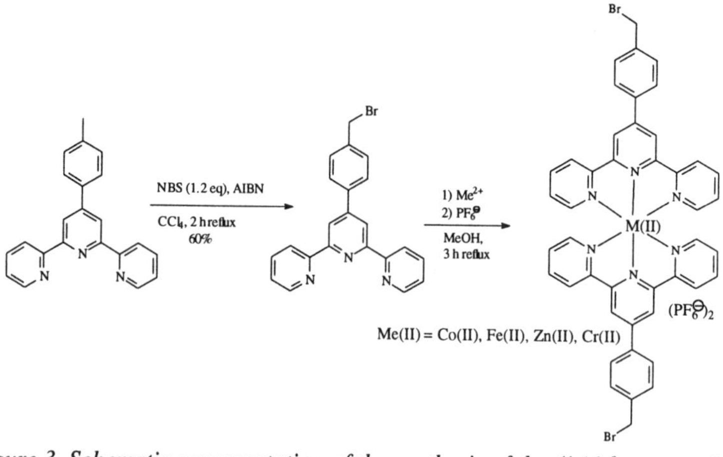

Figure 3. Schematic representation of the synthesis of the 4'-(4-bromomethyl-phenyl)-2,2':6',2"-terpyridine ligand and the corresponding metal complexes.

UV/VIS-spectroscopy demonstrates the specific differences in absorption of the various metal complexes (**Figure 4**). In particular, the iron complex shows a significant charge transition band at 568 nm, which is responsible for the deep purple color of the complex.

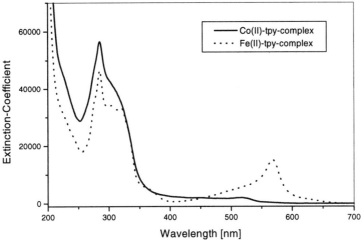

Figure 4. UV/VIS-spectra of different metallo-supramolecular initiator (in CH₃CN).

The metallo-supramolecular initiators were then stirred for one day with the 2-ethyl-2-oxazoline monomer in dry acetonitrile at 80°C. The resulting polymers were isolated after termination with piperidine or a piperidine functionalized terpyridine (for this strategy, see e.g. [23]) (**Figure 5**).

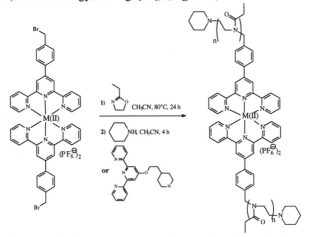

Figure 5. Schematic representation of the polymerization.

The metal-free functionalized ligands are also able to initiate the polymerization. However, we have demonstrated earlier that the uncomplexed bipyridine unit can act as termination agent resulting in branched polymers as well as in some crosslinked polymer chains [14]. The function of the metal ion during the polymerization process is thus similar to a classical protecting group which prevents the growing polymer chains from termination processes. Using this kind of initiator, poly(oxazoline)s with narrow and monomodal molar mass distributions between 1.06 and 1.25 could be obtained (**Figure 6**).

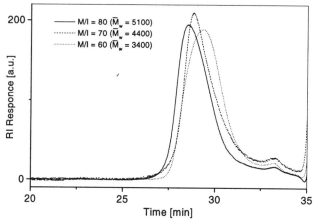

Figure 6. GPC curves of typical poly(ethyloxazoline)s with different M/I ratios. Initiated with the iron(II)-complex (CHCl₃ as eluent).

The molecular weight of these polymers can be controlled via the utilized monomer to initiator ratio (**Table 1, Figure 7**). Due to shear forces on the GPC a fragmentation of the supramolecular dimers can be observed (see also below).

Table 1. Poly(ethyloxazoline) polymerized utilizing iron(II)-metallo-supramolecular initiator (GPC data, polystyrene standards).

[M]/[I]	$\frac{1}{2}$ [M]/[I]	$\overline{M}_w$ (RI)	$\overline{M}_n$ (RI)	$\overline{M}_w / M_n$	calculated mass [g/mol]	(calculated mass)/2 [g/mol]
19	9.5	1510	1420	1.06	3041	1521
31	15.5	1510	1420	1.06	4229	2115
48	24	2340	2040	1.15	5912	2956
59	29.5	3650	2720	1.25	7000	3500
67	33.5	3560	2880	1.24	7793	3897
82	41	4600	3690	1.25	9278	4639

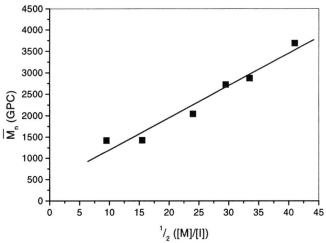

Figure 7. Dependence of the polymer mass on the [M]/[I] ratio (the average polymer masses are all about half the calculated values).

The central metal ion can be removed utilizing basic or acidic extraction conditions resulting in a terpyridine endfunctionalized uncomplexed polymer. This can be easily proven e.g. by UV/VIS spectroscopy experiments, NMR or AAS measurements (**Figure 8**).

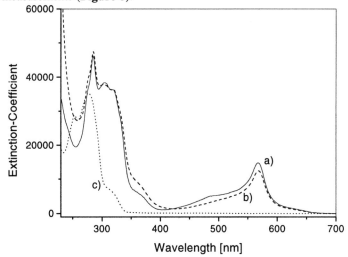

Figure 8. UV-VIS spectra of a poly(ethyloxazoline) utilizing the metallo-supramolecular initiator before (curve b) and after extraxction (curve c) as well as the initiator itself (curve a) (in CH₃CN).

Besides GPC measurements, MALDI-TOF-mass-spectrometry was used for the characterization of the complexed as well as of the decomplexed polymers (**Figure 9**). Each signal corresponds to a poly(oxazoline) with a terpyridine ligand or complex attached.

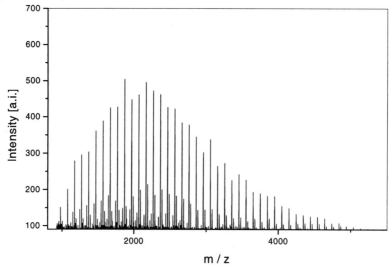

Figure 9. MALDI-TOF-MS of a decomplexed polymer with
$$\overline{M}_n(GPC) = 3450 \ g/mol, \ \overline{M}_w\big/\overline{M}_n = 1.19$$

Comparing the average molecular masses of the complexed and decomplexed polymers revealed not the presumed change of the molecular masses. As described above (**Table 1**) the GPC of the complexed polymers shows only the half of the calculated masses. After decomplexation the average polymer masses do not change very much but they constantly show slightly higher molecular masses. One possible explanation could be that the complexed polymers fragment by shear forces on the GPC column (see also [19]). Another reason could be that the hydrodynamic radius of the complexed poly(oxazoline)s differ strongly from the poly(styrene) standards. This phenomena is not yet completely understood but must be investigated in further studies.

The reversibility of the complexation/decomplexation process is another interesting question with important impact on potential applications (**Figure 10**).

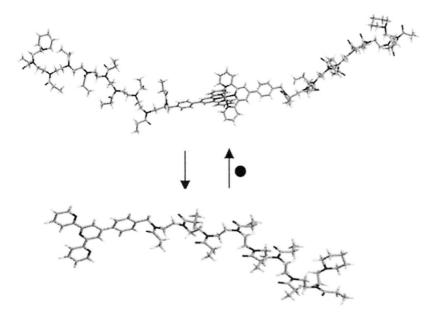

Figure 10. Model of the decomplexation process of the supramolecular poly(oxazoline)s.

To answer this question we performed titration experiments with the decomplexed polymer using an iron(II)-salt (iron(II)-sulfate heptahydrate) by stepwise increasing the concentration of the iron ions in the solution of a decomplexed polymer. For a quantitative expression we mixed equal volumes of the calibrated solution of the polymer and solutions of iron(II)-sulfate heptahydrate in different equivalent fractions (1/32; 1/16; 1/8; 1/4; 1/2) of the polymer concentration. Then we measured UV/VIS spectra of the mixtures (**Figure 11**). Looking at the charge transfer band at $\lambda = 568$ nm we observed an increase in absorption intensity with growing polymer/Fe(II) ratios. It could be shown, that up to 94% of the original extinction can be reached again by addition of Fe(II) salt, which indicates that the dimer is formed almost quantitatively, demonstrating the reversibility of the process.

Additional experiments showed, that it is also possible to polymerize 2-ethyloxazoline with other metal complexed initiators like Co(II)-complexes (**Figure 12**).

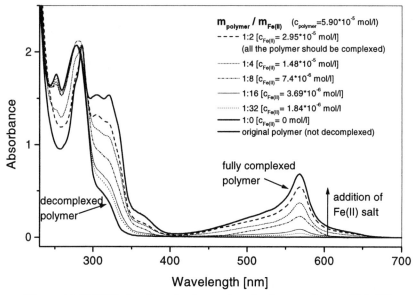

*Figure 11. UV/VIS-titration of a decomplexed polymer solution (CH₃CN)
with FeSO₄ × 7 H₂O (methanol).*

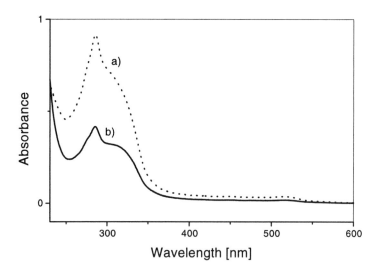

*Figure 12. UV/VIS-spectra of (a) a Co(II)-terpyridine-complex and (b) the
corresponding poly(oxazoline) complex (in acetone).*

Conclusion

The described results demonstrate a new entry towards functional polymers containing terpyridine segments. The supramolecular dimers with a metallo-supramolecular part can be manipulated by pH or shear forces and therefore the 3-dimensional structure can be controlled. The exact control of molecular weight, defined (functionalized) end groups and the variability of the polymer main chain itself offers an interesting combination between classical polymer chemistry and supramolecular science.

Acknowledgement: The research was supported by the *Bayerisches Staatsministerium für Unterricht, Kultus, Wissenschaft und Kunst*, the *Fonds der Chemischen Industrie*, the *Deutsche Forschungsgemeinschaft (DFG)* and the *BAYER AG*. We thank C. Eschbaumer for help with MALDI-TOF-MS and O. Nuyken for his support.

References

[1] Lehn, J.-M. *Supramolecular Chemistry - Concepts and Perspectives*, VCH, Weinheim, Germany, **1995**.

[2] Golden, J.H.; DeSalvo, F.J.; Fréchet, J.M.J.; Silcox, J.; Thomas, M.; Elman, J. *Science* **1996**, *273*, 782; Stupp, S.I.; LeBonheur, V.; Walker, K.; Li, L.S.; Huggins, K. E.; Keser, M.; Amstutz, A. *Science* **1997**, *276*, 384; Spatz, J.P.; Roescher, A.; Möller, M. *Adv. Mater.* **1996**, *8*, 337; Antonietti, M.; Henke, S.; Thünemann, A. *Adv. Mater.* **1996**, *8*, 41.

[3] Xia, Y.; McClelland, J.J.; Gupta, R.; Qin, D.; Yhao, X.-M.; Sohn, L.L.; Celotta, R.J.; Whitesides, G.M. *Adv. Mater.* **1997**, *9*, 147; Renak, M.L.; Bazan, G.C.; Roitman, D. *Adv. Mater.* **1997**, *9*, 392.

[4] Ohno, H.; Mukaigawa, M. *J. Electronal. Chem.* **1998**, *452*, 141; Neve, F.; Crispini, A.; Campagna, S.; Serroni, S. *Inorg. Chem.* **1999**, *38*, 2250; Collin, J.P.; Guillerez, S.; Sauvage, J.P.; Barigelleti, F.; De Cola, L.; Flamigni, L.; Balzani, V. *Inorg. Chem.* **1991**, *30*, 4230; Collin, J.P.; Jouaiti, A.; Sauvage, J.P. *J. Electroanal. Chem. Interfacial Electrochem.* **1990**, *75*, 286.

[5] Bonhote, P.; Moser, J.-E., Humphry-Baker, R.; Vlachopoulos, N.; Zakeeruddin, S.M.; Walder, L.; Graetzel, M. *J. Am. Chem. Soc.* **1999**, *121*, 1324; Thompson, A.M.W.C.; Smailes, M.C.C.; Jeffery, J.C.; Ward, M.D. *J. Chem. Soc., Dalton Trans.* **1997**, 737; Ohno, H.; Mukaigawa, M.J. *Kidouri* **1998**, *32*, 24; Ng, P.K.; Ng, W.Y.; Gong, X.; Chan, W.K. **1998**, *488*, 581.

176

[6]   Farrer, B.T.; Thorp, H.H. *Inorg. Chem.* **2000**, *39*, 44; Barrios, A.M.;
      Lippard, S.J. *J. Am. Chem. Soc.* **1999**, *121*, 11751; Carter, P.J.; Cheng, C.-
      C.; Thorp, H.H. *J. Am. Chem. Soc.* **1998**, *120*, 632; Suh; J.; Hong, S.H. *J.
      Am. Chem. Soc.* **1998**, *120*, 12545.
[7]   Constable, E.C.; Mundwiler, S. *Polyhedron* **1999**, *18*, 2433; Weidener, S.;
      Pikramenou, Z. *Chem. Commun.* **1998**, 1473; Constable; E.C.; Fallahpour,
      R.-A. *J. Chem. Soc., Dalton Trans.* **1996**, 2389.
[8]   B. Hasenknopf: *Dissertation*, Université Strasbourg, **1996**.
[9]   Hanabusa, K.; Nakamura, A.; Koyama, T.; Shirai, H. *Makromol. Chem.*
      1992, *193*, 1309.
[10] Spahni, W.; Calzaferri, G. *Helv. Chim. Acta* **1984**, *67*, 450.
[11] Schubert, U.S.; Hochwimmer, G. *Polym. Preprints* **1999**, *40(2)*, 340.
[12] Schubert, U.S.; Hochwimmer, G. *Polym. Preprints* **2000**, *41(1)*, 433.
[13] Kobayashi, S. *Progr. Polym. Sci.* **1990**, *15*, 751; Chujo, Y.; Saegusa, T. in
      "Ring Opening Polymerization", Ed. Brunelle, D.I., Hanser, München,
      1993.
[14] Hochwimmer, G.; Nuyken, O.; Schubert, U.S. *Macromol. Rapid Commun.*
      1998, *19*, 309.
[15] Schubert, U.S.; Nuyken, O.; Hochwimmer, G. *Division of Polymeric
      Materials: Science and Engineering, Preprints* **1999**, *80*, 193; Schubert,
      U.S.; Eschbaumer, C.; Hochwimmer, G. *Tetrahedron Lett.* **1998**, *39*, 8643.
[16] Lamba, J.J.S.; Fraser, C.L. *J. Am. Chem. Soc.* **1997**, *119*, 1801; Lamba,
      J.J.S.; McAlvin, J.E.; Peters, B.P.; Fraser, C.L. *Polym. Preprints* **1997**,
      38(1), 193; Collins, J.E.; Fraser, C.L. *Macromolecules* **1998**, *31*, 6715.
[17] Schubert, U.S.; Hochwimmer, G. *Polym. Preprints* **1999**, *40(2)*, 1068.
[18] McWhinnie, W.R.; Miller, J.D. *Adv. Inorg. Chem. Radiochem.* **1969**, *12*,
      135; Constable, E.C. *Adv. Inorg. Chem. Radiochem.* **1986**, *30*, 69.
[19] Schubert, U.S.; Eschbaumer, C.; Nuyken, O.; Hochwimmer, G. *J. Incl.
      Phenom.* **1999**, *32*, 23.
[20] Schubert, U.S.; Hochwimmer, G.; Heller, M., in preparation.
[21] Bushell, K.L.; Couchman, S.M.; Jeffery, J.C.; Rees, L.H.; Ward, M.D. *J.
      Chem. Soc., Dalton Trans* **1998**, 3397; Collin, J.P.; Heitz, V.; Sauvage, J.P.
      Tetrahedron Lett. **1991**, *32*, 5977.
[22] Armspach, D.; Constable, E.C.; Housecroft, C.E.; Neuburger, M.; Zehnder,
      M. *J. Organomet. Chem.* **1998**, *550*, 193; Hanabusa, K.; Nakamura, A.;
      Koyama, T.; Shirai, H. *Polym. Int.* **1994**, *35*, 231; Whittle, B.; Batten, S.R.;
      Jeffery, J.C.; Rees, L.H.; Ward, M.D. *J. Chem. Soc., Dalton Trans.* **1996**,
      22, 4249; Collin, J.P.; Guillerez, S.; Sauvage, J.P. *J. Chem. Soc., Chem.
      Commun.* **1989**, 776.
[23] Nuyken, O.; Maier, G.; Groß, A.; Fischer, H. *Macromol. Chem. Phys.*
      1996, *197*, 83.

Chapter 13

Reversed-Type Micelle Formation Property of End-Glycosidated Polystyrene

Toyoji Kakuchi and Naoya Sugimoto

Division of Bioscience, Graduate School of Environmental Earth Science, Hokkaido University, Sapporo 060-0810, Japan

The controlled radical polymerization of styrene using the TEMPO-based initiators with carbohydrate residues is a novel method for producing end-glycosidated polystyrenes. Xylose, glucose, galactose, cellobiose, maltose, maltotriose, maltotetraose, maltopentaose, and maltohexaose were used as the carbohydrate residues. The well-defined end-glycosidated polystyrene aggregated to form the reversed-type micelle in benzene solution. The aggregation number varied from 2.2 to 62.5 depending on the polystyrene chain length and the number of pyranose units.

Macromolecular design and the synthesis of polymers consisting of hydrophilic and hydrophobic units have drawn attention from the viewpoint of constructing synthetic macromolecules with a higher structural order. Thus, there are many efforts to synthesize various types of well-defined polymers using carbohydrates as hydrophilic units, because carbohydrates, compounds with hydroxy groups, are easily available raw materials. In addition, living polymerization techniques are used for the synthesis of carbohydrate-containing polymers with well-defined structures. For example, ROMP (*1*) and living cationic polymerization (*2*) techniques using appropriate monomers were performed to afford homo- and block copolymers having carbohydrate residues as side chains, and living anionic polymerization methods combined with terminators (*3*) or initiators (*4*) possessing carbohydrate residues produced end-glycosidated polymers.

Recently, we reported the synthetic method for the well-defined end-functionalized polymers with mono- and disaccharide residues using controlled radical polymerization (*5*). Sufficient control of the molecular weight and molecular weight distribution as well as the end-functionality was realized in each of the polymer samples synthesised using TEMPO-based initiators having acetylated carbohydrate residues. Thus, of great interest is studying the micelle formation properties of end-glycosidated polystyrenes. In this chapter, we describe the synthesis of well-defined end-functionalized polystyrenes (**3**) by the polymerization of styrene using the TEMPO-based initiators (**1**) having acetylated carbohydrate residues, such as xylose (**a**), glucose (**b**), galactose (**c**), cellobiose (**d**), maltose (**e**), maltotriose (**f**), maltotetraose (**g**), maltopentaose (**h**), and maltohexaose (**i**). In addition, the effect of polystyrene chain length and type of end-carbohydrate groups on the aggregation number for the micelle of **3** in benzene solution was investigated.

Synthesis of End-glycosidated Polystyrene

Georges *et al.* reported that narrow molecular weight polymers can be prepared by the radical polymerization of styrene using a traditional radical initiator, benzoyl peroxide and 2,2,6,6-tetramethylpiperidinyloxyl (TEMPO) as the stable free radical (*6*). For the "living" radical polymerization, the adduct of styrene and TEMPO is isolated as a stable compound, and many TEMPO-based initiators have been used for the synthesis of various macromolecular architectures. Thus the controlled radical polymerization method can be used to obtain carbohydrate residues as end-functional groups.

The TEMPO-based initiators having xylose (**1a**), glucose (**1b**), galactose (**1c**), cellobiose (**1d**), maltose (**1e**), maltotriose (**1f**), maltotetraose (**1g**), maltopentaose (**1h**), and maltohexaose (**1i**) were synthesized by the reaction of 4-ethylphenyl derivatives of the corresponding carbohydrates with TEMPO in the presence of di-*tert*-butyl peroxalate (*7*).

The bulk polymerization of styrene initiated by the TEMPO-based initiators was carried out at 120 °C for 6 h. Table 1 summarizes the typical results using initiators **1b** and **1e**. The polymer yield was 35 ~ 52 % for **2b** and 38 ~ 45 % for **2e**. The value of M_n increased with increasing molar ratio of styrene and initiator ([styrene]/[initiator]) from 10,500 to 31,300 for **2b** and from 12,700 to 30,700

Scheme 1

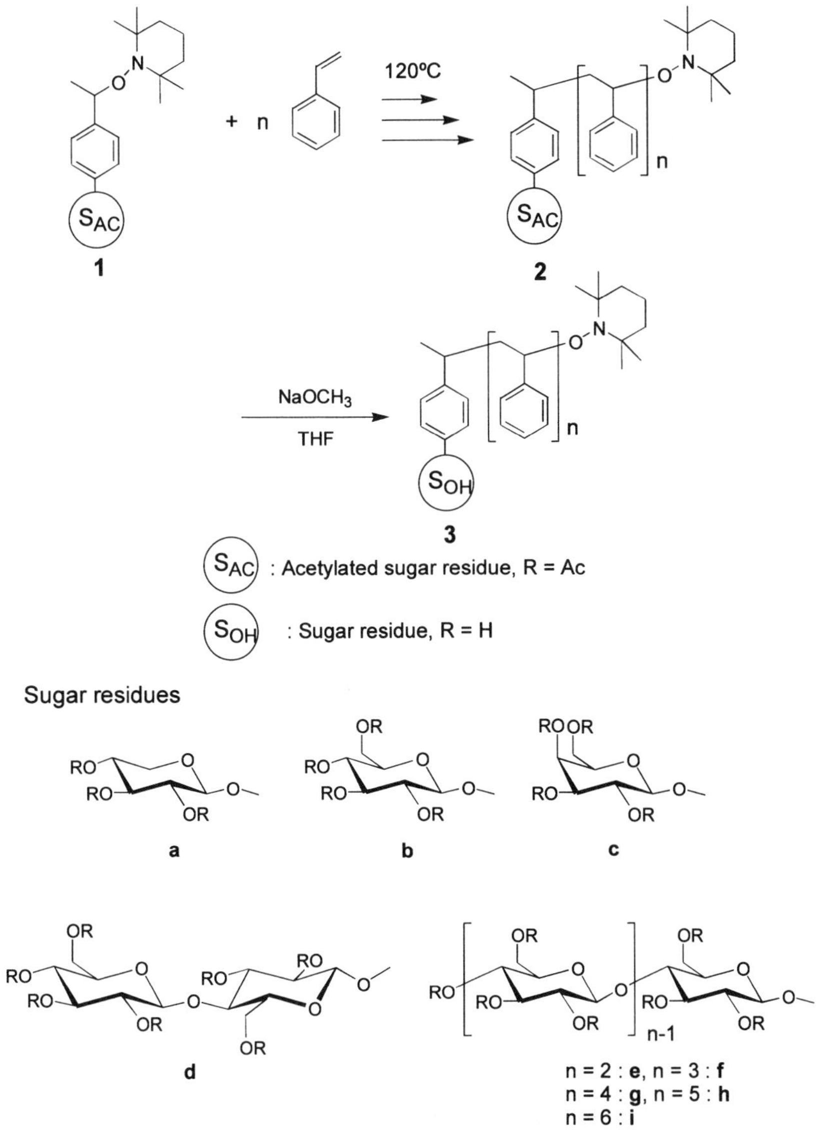

S_{AC} : Acetylated sugar residue, R = Ac

S_{OH} : Sugar residue, R = H

Sugar residues

a

b

c

d

n = 2 : **e**, n = 3 : **f**
n = 4 : **g**, n = 5 : **h**
n = 6 : **i**

Table 1. Bulk polymerization of styrene using initiators 1 and 2 [a]

initiator	$\dfrac{[styrene]}{[nitiator]}$	yield (%)	M_n [b]	M_w/M_n [b]
3b	300	35	10,500	1.10
3b	600	44	20,600	1.12
3b	900	52	31,300	1.14
3e	300	38	12,700	1.10
3e	600	45	23,700	1.13
3e	900	42	30,700	1.15

a) Temp., 120 °C; time, 6 h.

b) Determined by GPC using a polystyrene standard.value

for **2e**, and in all cases the polydispersity (M_w/M_n) was very narrow, *i.e.*, the value of M_w/M_n is 1.10 ~ 1.15. Figure 1 shows the polymer yields for the bulk polymerization of styrene using initiators **1b** and **1e** ([styrene]/[initiator] = 600). Both systems display similar polymerization rates. The molecular weight of the polymers was found to increase in an approximately linear fashion with conversions, while low M_w/M_ns, such as 1.11 ~ 1.13 are maintained throughout the polymerization for both systems. For the polymerization with [styrene]/[initiator] = 600 at 120 °C for 3 h, the polymer yield and the M_n (M_w/M_n) were 28 % and 14,300 (1.17) for **1a**, 33 % and 12,800 (1.20) for **1c**, 29 % and 18,000 (1.12) for **1e**, 25 % and 12,500 (1.18) for **1f**, and 24 % and

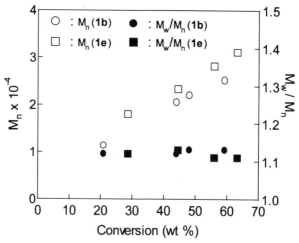

Figure 1. Evolution of experimental molecular weight and polydispersity with conversion in the bulk polymerization of styrene initiated by initiators **1b** and **1e** .[styrene]/[initiator] = 600, temp. = 120 °C.

20,900 (1.21) for **1i**. These results indicated that the TEMPO-based initiator was effective for the controlled radical polymerization leading to well-defined polymers.

In the ^{1}H NMR spectra of **2b** and **2e**, the signals at 5.3 ~ 3.9 ppm for **2b** and 5.5 ~ 3.8 ppm for **2e** are assigned to the glucose and maltose units, respectively. The degrees of end-functionalization, which were determined by the ^{1}H NMR peak intensities, were nearly quantitative, *i.e.*, ca. 1.0. It was observed that end-functionalized poly(styrene)s with glucose and maltose residues were completely separated from the unfunctionalized polystyrene by TLC using an SiO$_2$ plate with toluene as the mobile phase. Under the developed conditions, polymers **2b** and **2e** always gave spots that remained near the spotting points (R_f = ca. 0), whereas unfunctionalized polystyrene was quite mobile to near the top (R_f = 0.9). Since the amount of each spot can be quantitatively detected by a flame ionization detector (FID), the degrees of end-functionalization can be determined by comparing each peak area. Accordingly, the TLC method coupled with FID becomes very effective for the quantitative analysis of the end-functionalized polymers, in which the experimental error is as much as 3 wt %. The degrees of end-functionalization by the analysis of TLC-FID method are also indicated to be nearly quantitative for **2b** and **2e**. Deacetylation of **2** was performed with sodium methoxide in THF to quantitatively afford the end-glycosidated polystyrene, **3**.

These analytical results suggested that the polymerization of styrene using

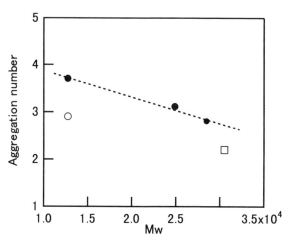

Figure 2. Aggregation number of end-glycosidated polystyrenes in benzene as a function of weight average molar mass. **3a** (○), **3b** (●), **3c** (□)

the TEMPO-based initiators with carbohydrate residues proceeded satisfactorily to afford the well-defined end-glycosidated polystyrenes.

Reversed-type Micelle Formation of End-glycosidated Polystyrenes

Recently, block copolymers and end-functionalized polymers have been studied in terms of their phase separation and self-assembly abilities. Eisenberg et al. reported that sodium carboxylates of hydrophilic terminal groups were separated from the hydrophobic polystyrene main chains to form reverse-type micelles in cyclohexane by aggregating several polymer chains (7). In addition, Hirao et al. reported that polystyrene with one and two glucose residues at their chain ends were aggregated to form reverse-type micelles in cyclohexane (3). Thus, for the end-glycosidated polystyrene, of great interest is to study the effect of the number of pyranose units on the reverse-type micelle formation properties.

For the ^{1}H NMR spectra of end-glycosidated polystyrene, 3, in C_6D_6, it was difficult to observe the carbohydrate units in 3, suggesting the aggregation of 3 leading to reverse-type micelle formation. Thus, the micelle formation of end-glycosidated polystyrenes was observed by laser light scattering measurements in benzene solution. Figures 2 and 3 show the plots of the aggregation numbers vs. molecular weights of the end-functionalized polystyrenes with glucose and maltose residues (3b and 3e, respectively). The aggregation number decreased with increasing the molecular weight, i.e., from 3.7 to 3.0 with increasing Mw from 12,800 to 28,500 for 3b and from 9.1 to 3.7 with increasing Mw from

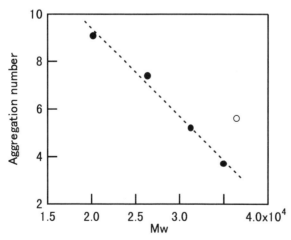

Figure 3. Aggregation number of end-glycosidated polystyrenes in benzene as a function of weight average molar mass. **3d** (○), **3e** (●)

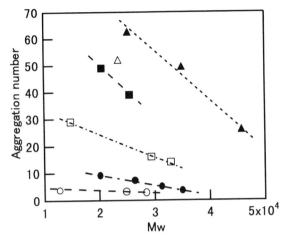

Figure 4. Aggregation number of end-glycosidated polystyrenes in benzene as a function of weight average molar mass. **3b**(○), **3e** (●), **3f** (□) **3g** (■), **3h** (△), **3i** (▲)

20,200 to 35,000 for **3e**. For the end-fuctionalized polymer with a monosaccharide, the micelle formation property of the polymers with xylose and galactose residues, **3a** and **3c**, slightly differed from that of **3b**, because the aggregation numbers for **3a** and **3c** deviated from the straight line for **3b**, as shown in Figure 4. A similar tendency for the end-fuctionalized polymers with the disaccharides, **3d** and **3e**, was also observed in Figure 3. These results indicated that micelle formation property of the end-glycosidated polystyrene may be affected by the type of carbohydrate residues.

Figure 4 shows the plots of the aggregation numbers vs. molecular weights of the end-functionalized polystyrenes with mono-, di-, and oligosaccharides, i.e., glucose (**3b**), maltose (**3e**), maltotriose (**3f**), maltotetraose (**3g**), maltopentaose (**3h**), and maltohexaose (**3i**). For all the end-glycosidated polystyrene, the aggregation number decreased with increasing molecular weight, e.g., from 62.5 to 26.0 with increasing Mw from 20,900 to 45,800. In addition, the aggregation number increased with the increasing number of pyranose units, i.e., the aggregation number increased in the order of glucose (**3b**) > maltose (**3e**) > maltotriose (**3f**) > maltotetraose (**3g**) > maltopentaose (**3h**) > maltohexaose (**3i**). These results indicated that changing the polystyrene chain length and the number of pyranose units should control the micelle size of the end-glycosidated polystyrenes.

Experimental Part

Materials and measurements. The molecular weights of the resulting polymers were measured in chloroform using a Jasco GPC-900 gel permeation chromatograph equipped with two polystyrene gel columns (TSKgel GMHHR-M, TOSOH Corporation). The number-average molecular weight (M_n) and molecular weight distribution (M_w/M_n) of the polymers were calculated on the basis of polystyrene calibration. The TLC-FID measurements were carried out using an IATRON MK-5 (Iatron Laboratories, Inc.). Laser light scattering measurements were performed with a Super-Dynamic Light Scattering Spectrophotometer DLS-7000AL (Ohtsuka Electronics Co., Ltd.) in benzene. All the polymer solutions were clarified using a 0.1 μm millipore filter.

Polymerization. A mixture of **1b** (29.2 mg, 0.048 mmol) and styrene (3.0 g, 28.8 mmol) was heated at 120 °C for 15 h under a N_2 atmosphere. After cooling, the reaction mixture was diluted with $CHCl_3$, and then poured into methanol (300 mL) after which the precipitate was filtered off. The obtained precipitate was purified by reprecipitation in chloroform-methanol and dried *in vacuo* to yield the polymer (1.82 g, 59 %). The M_n and M_w/M_n were 25,200 and 1.13, respectively.

Deacetylation. After a mixture of 0.5 g of **2b** ($M_n = 20,600$) and 2[M] sodium methoxide (5 mL) in 15 mL of dry THF was stirred at room temperature for 12 h, the whole solution was poured into 200mL of methanol/water (v/v, 9:1). The precipitate was filtered off, washed with water and then methanol, and purified by two reprecipitations from chloroform/methanol. After freeze-drying of a benzene solution of the product (**3b**), a white powder was quantitatively obtained.

References

1. Nomura, K.; Schrock, R. R. *Macromolecules,* **1996**, *29,* 540.
2. Yamada, K.; Yamaoka, K.; Minoda, M.; Miyamoto, T. *J. Polym. Sci.: Part A: Polym. Chem.,* **1997**, *35,* 255.
3. Hayashi, M.; Loykulnant, S.; Hirao, A.; Nakahama, S. *Macromolecules,* **1998**, *31,* 2057.
4. Aoi, K.; Suzuki, H.; Okada, M. *Macromolecules,* **1992**, *25,* 7073.
5. Sugimoto, N.; Kakuchi, T. *Polymer Preprints,* **1999**, *40,* 111.
6. Georges, M. K.; Veregin R. P. N. *Macromolecules,* **1993**, *26,* 2987.
7. Yamada, B. *Macromolecules,* **1998**, *31,* 4659.
8. Desjardins, A.; Theo, G. M.; Eisenberg, A. *Macromolecules,* **1992**, *25,* 2412.

Chapter 14

Ultracentrifugation Studies on the Solution Properties of Supramolecular Building Blocks for Polymers: Potential, Problems, and Solutions

C. Tziatzios[1], H. Durchschlag[2], C. H. Weidl[3], C. Eschbaumer[3], W. Maechtle[4], P. Schuck[5], U. S. Schubert[3], and D. Schubert[1,*]

[1]Institut fuer Biophysik, JWG-Universitaet, D-60590 Frankfurt am Main, Germany
[2]Institut fuer Biophysik und Physikalische Biochemie, Universitaet Regensburg, D-93040 Regensburg, Germany
[3]Lehrstuhl fuer Makromolekulare Stoffe, Technische Universitaet Muenchen, D-85747 Garching, Germany
[4]Kunststofflaboratorium, BASF AG, D-67056 Ludwigshafen, Germany
[5]Molecular Interactions Resource, DBEPS, ORS, National Institutes of Health, Bethesda, MD 20892-5766

Suitable supramolecular building blocks can assemble via non-covalent interactions into large, often regular structures, such as surface films. Uniformity of the state of association of the building blocks seems to be a prerequisite for the success of the assembly process. The paper shows how the association behavior of the compounds can be characterized by analytical ultracentrifugation. In addition to describing the potential of the method, two major problems are addressed: the determination of the partial specific volume of the compounds, $\bar{v}$, and disturbancies due to their electric charge. $\bar{v}$ is found to be strongly dependent on the solvent, but can be reliably determined by a special solvent density variation technique ("buoyant density method"). The latter procedure also seems to correct for the secondary charge effect.

Supramolecular chemistry has its own approach towards producing large, polymer-like structures. In this approach, pseudo-polymers are build up in a two-step procedure, via non-covalent associations. In the first step, relatively small units (in most cases with molar masses of a few hundred g/mol) are designed to assemble spontaneously via metal coordination interactions, ionic interactions or hydrogen bonds into supramolecular complexes (*1,2*). In a second step, these complexes further associate into extended, frequently regular structures, such as surface films (*3*).

The success of the second self-assembly process leading from the supramolecular complex to the polymer is by no means granted. It relies not only on the appropriate structure of the supramolecular "monomer", but also depends strongly on its state of association in solution: It seems to require a uniform, preferentially monomeric starting material. This, in turn requires a careful control of the state of association of the supramolecular compounds in solution and, in parallel, variations in the experimental parameters until monodispersity of the building blocks is achieved. In 1997, our group has suggested that analytical ultracentrifugation should be the method of choice for characterizing the association behavior of the supramolecular compounds in solution, and we have shown that application of the technique to this problem is feasible (*4*). More recently, we have reviewed the available methods (*5*) and reported a detailed application to a special class of supramolecular compounds, namely metal coordination arrays (*6*). We have also shown that monomeric arrays can serve as building blocks for the formation of extended, very regular surface films (*3*). In ref. (*6*), special attention was given to the probably most critical problem in analytical ultracentrifugation of supramolecular compounds, the determination of their "partial specific volume" $\bar{v}$. In the present paper, we have extended these studies, with special emphasis on the "buoyant density method", again using a metal coordination array as a model compound. We will show that the $\bar{v}$-problem is even more critical than anticipated, but we will also present solutions to the problem. In addition, we will focus on problems inherent to analytical ultracentrifugation of electrically charged solutes, the so-called primary and secondary charge effects.

Principles of Analytical Ultracentrifugation

Analytical ultracentrifugation is the study of the movement or the distribution of solutes under the influence of a centrifugal force. In most cases the measurements are performed in an instrument based on a standard preparative ultracentrifuge. Added are a special rotor, special cells equipped with quartz or sapphire windows, and an optical system for measuring light absorbance or refractive index changes in the ultracentrifuge cell as a function of the distance from the axis of rotation, r. The most important types of experiments, which have recently been reviewed in more detail (*5*), are the following:

- Sedimentation velocity runs: Here, the time course of the local concentration distribution of the sedimenting molecules, c(r,t), is used to extract the "sedimentation coefficient", s, the diffusion coefficient, D, and from them, applying the "Svedberg equation", the effective or buoyant molar mass of the particles, $M_{eff} = M(1 - \bar{v}\rho_o)$, where $\bar{v}$ denotes the partial specific volume of the molecules (virtually equal to their reciprocal density) and ρ_o the density of the solvent. The duration of the experiments is a few hours. For modern data analysis see (7). The analysis can resolve up to 2-3 different sedimenting species.
- Archibald runs: A rough estimate ($\pm$ 10-15 %) of the average buoyant molar mass of the sedimenting species is obtained from the initial phase (30-60 min) of a sedimentation equilibrium run, using c(r,t)-data only from the region near the meniscus of the ultracentrifuge cell (8).
- Sedimentation equilibrium runs: In these experiments, centrifugation is performed until, after several hours up to a few days, a time-independent distribution of the sample molecules is attained. For uniform particles, the equilibrium distribution, c(r), can be described by a Boltzmann distribution:

$$c(r) = c(r_o)\exp\left[\frac{M(1-\bar{v}\rho_o)\omega^2}{2RT}(r^2 - r_o^2)\right], \qquad (1)$$

where r_o denotes an arbitrary reference radius and ω the angular velocity of the rotor. The corresponding description for mixtures of n different particles is a sum of Boltzmann distributions:

$$c(r) = \sum_{k=1}^{n} c_k(r_o)\exp\left[\frac{M_k(1-\bar{v}_k\rho_o)\omega^2}{2RT}(r^2 - r_o^2)\right] \qquad (2)$$

Sedimentation equilibrium analysis is the method of choice for analyzing association equilibria. Generally, it can resolve up to 3-4 different species.

The foregoing description is based on the assumption that the sedimenting particles show "ideal" sedimentation behavior. To achieve this, particle concentration has to be low enough (in practice: < 10 μM). An additional prerequisite with charged solutes is that the solutions should contain some supporting electrolyte (see below). For detailed general treatments of the technique, the reader is referred to two classical monographs (9,10).

Materials and Methods

The metal coordination array studied (compound f of (6); Figure 1) was prepared as described in (11). All other chemicals were purchased from Merck (Darmstadt) or Sigma/Aldrich (Deisenhofen) and were of analytical or spectroscopic grade (if available).

Sedimentation equilibrium experiments were performed in a Beckman Optima XL-A ultracentrifuge, using an An-Ti 50 rotor, titanium double sector centerpieces of 12 mm pathlength (BASF) and polyethylene gaskets (4). Sample volume was 200 µl or, in a few control experiments which aimed at detecting possible pressure effects (9,10), 100 µl. Rotor speed was 40,000 rpm, rotor temperature 20 ^{0}C. Compound concentration was approx. 18 µM. The absorbance-versus-radius profiles, A(r), were recorded at the wavelength of an absorbance maximum of the cobalt coordination array, 330 nm. The evaluation of the data used the computer program DISCREEQ by P. Schuck (12,13). The determination of the baseline position followed (4). M_{eff}-values were determined from one-component fits (eq. (1)). The densities of solvents and solutions were measured with a Paar DMA 02 digital density meter.

Application to a Supramolecular Compound: Problems and Solutions

The $\bar{v}$-Problem

We have recently compared different methods for determining the partial specific volume, $\bar{v}$, of a number of different metal coordination arrays (6). Some of these compounds were demonstrated to represent suitable starting materials for the preparation of extended surface layers (3). We have shown that $\bar{v}$ is the most critical parameter in the description of the compounds' association behavior. It was found that all methods applied by us suffered from severe problems:

1. Calculating $\bar{v}$ from increments, as routinely done with proteins, organic compounds, and polymers in aqueous solution (14,15), is hampered by the unknown behavior of the building blocks in organic solvents.
2. Determination by digital densimetry, the standard experimental technique (14), can lead to highly erroneous results due to the formation of microcrystals or clusters of the solutes at the high compound concentrations required. A lack of material frequently represents another serious obstacle.
3. The two versions of the Edelstein-Schachman procedure, in which M and $\bar{v}$ are determined simultaneously from the observed buoyant molar mass values, M_{eff}, measured in two solvents of different density, also are of

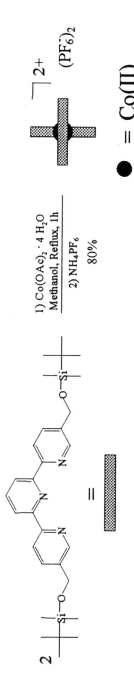

Figure 1. Assembly and structure of the metal coordination array used as a model system. Two of the organic ligands plus one cobalt(II) ion build up, by self-assembly, the grid-like structure shown at the right. Its molar mass (including two PF_6^- counterions) is 1393 g/mol.

limited applicability: The original procedure (*16*), which uses the nondeuterated and the deuterated forms of the same solvent, has a low accuracy, due to the smallness of the density differences; the modified version, applying two different solvents (*6*), has to assume that $\bar{v}$ is solvent-independent. In addition, both procedures will work only if the compound is monomeric or aggregated homogeneously and in the same way in the two solvents.

4. The buoyant density method, which uses solvent mixtures to determine $\bar{v}$ as the reciprocal of that solvent density for which the compound is neutrally buoyant ($M_{eff} = 0$), also is based on the assumption that $\bar{v}$ is solvent-independent. With respect to this assumption we have shown that it is invalid, the demonstrated differences in $\bar{v}$ between two different solvent systems amounting to 7 % (*6,17*).

Comparing the two ultracentrifuge-based methods, 3. and 4., the latter method has the great advantage that a moderate heterogeneity of the material (including solvent-dependent variations) should have little or no influence on $\bar{v}$, since oligomers from the same size range are thought to behave similarly (*14,15*). In addition, nonideal sedimentation behavior of the sample due to the primary charge effect (see below) will not affect the $M_{eff} = 0$-position on the ρ_o-scale. Since the common solvents cover a large range of solvent densities, most supramolecular compounds can be studied this way. In view of these principal advantages of the method, we have explored whether the solvent-dependency of $\bar{v}$ could be suppressed. Since, according to the data of ref. (*6*), solvent polarity seems to play an important role, we have tried to overcome this effect by adding supporting electrolyte (this will also suppress the primary charge effect, see below).

As a model system for our studies aiming at establishing the buoyant density method, we chose the compound of Figure 1, the most intensely studied compound in our earlier paper (*6*). Figure 2 shows apparent (average) M_{eff}-data as a function of solvent density, collected in different solvent mixtures and in the absence and presence of salt. It is obvious from the figure that the presence of salt has a large effect. However, the magnitude and even the direction of the $\bar{v}$-change depend on the solvent system. In all cases studied, the effect showed saturation at salt concentrations exceeding a few mmoles/l (Figure 3). The total variation in $\bar{v}$ under the conditions tested was approx. 12 % of its maximum value. Thus, a unique value for $\bar{v}$, applicable to a large number of different organic solvents and solvent mixtures and salt concentrations, apparently does not exist. Instead, $\bar{v}$ has to be determined separately for each solvent condition. (A dependency of $\bar{v}$ on the solvent was already described for synthetic polymers (*18*)).

To test whether $\bar{v}$ could also be influenced by pressure effects (9,10), we have compared results obtained at different sample volumes (200 vs. 100 µl).

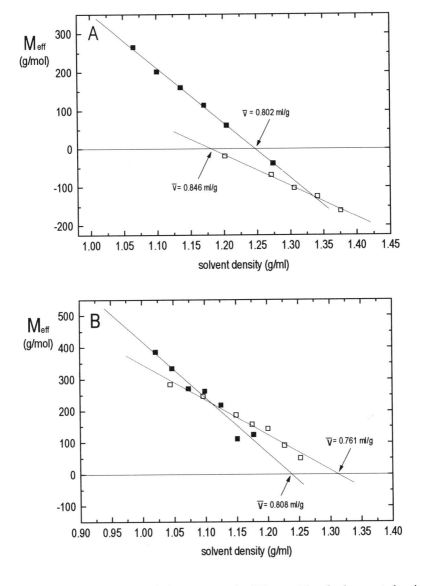

Figure 2. Determining $\bar{v}$ of the compound of Figure 1 by the buoyant density method: Dependency of M_{eff}, as obtained by sedimentation equilibrium analysis, on solvent density adjusted by different solvent mixtures, in the absence (□) or presence (■) of 25 mM NH_4PF_6. Solvent systems: (A) acetonitrile/chloroform; (B) dichloroethane/anisole.

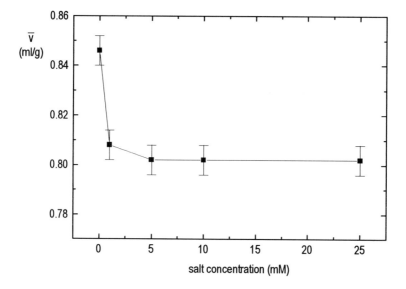

Figure 3. Dependency of $\bar{v}$, as determined from plots according to Figure 2, on the concentration of added NH_4PF_6 in acetonitrile/chloroform mixtures.

We found that, in these experiments, both the M_{eff}-values and the quality of the fits were indistinguishable. Thus, under the conditions employed $\bar{v}$ does not critically depend on hydrostatic pressure.

How can, in view of the complex dependencies described, $\bar{v}$ and subsequently M be determined reliably? We think that $M_{eff}(\rho_o)$-plots analogous to Figure 2, measured in the presence of salt concentrations high enough to suppress nonideal sedimentation behavior (see below), still represent the best possible approach. However, two prerequisites should be fulfilled: (i) The $M_{eff}(\rho_o)$-plot should preferentially be linear over an extended ρ_o-range including the zero crossover (in this case, the constancy of the slope, which corresponds to $-M\bar{v}$, normally will indicate constancy of both parameters under the conditions applied). (ii) The experimental condition for which M (or the state of association) is to be determined should be covered by these data; in addition, the corresponding M_{eff} should be precise enough (i.e., far from the zero crossover of $M_{eff}(\rho_o)$).

It should be noted that even a nonlinear $M_{eff}(\rho_o)$-plot may yield correct figures for both M and the corresponding $\bar{v}$; however, to proof this may be difficult. On the other hand, an apparent linearity could also result from adverse changes in M or $\bar{v}$ or both of them. Thus, each analysis applying the procedure suggested above has to be preceded by a critical examination of the experimental data. Plausibility of the slope of the straight line in relation to its nominal value, $-M\bar{v}$, could be a helpful criterion.

The application of the above suggestions to two data sets on the cobalt grid of Figure 1 is demonstrated in Figures 4 and 5. Here, we attempted to determine M (or the state of association) of the compound in solution of acetonitrile plus 25 mM NH_4PF_6 as supporting electrolyte (see below). In order to determine the corresponding $\bar{v}$, the solvent was mixed with a more dense one, either DMSO or chloroform. The $M_{eff}(\rho_o)$-plots are shown in Figure 4. With the acetonitrile/DMSO mixture, the plot is linear over the accessible ρ_o-range, and the (extrapolated) zero crossover position yields $\bar{v} = (0.854 \pm 0.008)$ ml/g. Together with $M_{eff} = 466$ g/mol in acetonitrile and $\rho_o = 0.7885$ g/ml, M in this solvent is found as (1437 ± 60) g/mol. The molar mass value includes two PF_6 counteranions, which form a part of the electroneutral solute component and cosediment with the compound (19). It agrees with the theoretical value M=1393 g/mol for the monomeric grid within the precision of the measurement. - The corresponding fit to the A(r)-data is of excellent quality, as shown in Figure 5. An alternative fit based on a monomer/dimer model yields a dimer content of virtually zero (see also below).

The data just discussed indicate that the compound studied is monomeric both in acetonitrile and in the acetonitrile/DMSO mixtures used. In solutions with a density close to $\bar{v}^{-1}$, i.e. at higher DMSO content, a slight self-association would, however, most probably remain undetected due to the small slope of the A(r)-curve). On the other hand, the data collected with acetonitrile/chloroform

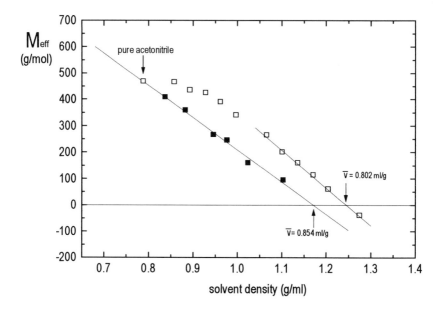

Figure 4. M_{eff} (ρ_o) for the cobalt grid over an extended ρ_o-range. Solvent systems: (■) acetonitrile/DMSO; (□) acetonitrile/chloroform. The solvents contained 25 mM NH_4PF_6.

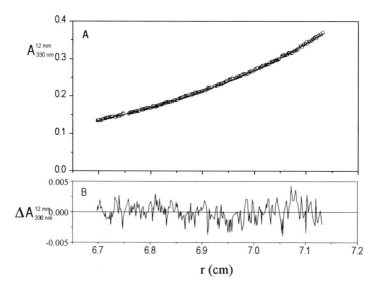

Figure 5. (A) Best one-component fit (—) to the experimental A(r)-data (O) for the cobalt coordination array in acetonitrile plus supporting electrolyte, based on $\overline{v} = 0.854$ ml/g and an optimum molar mass of 1437 g/mol. (B) Residuals $\Delta A(r)$ of the fit.

mixtures are indicative of a more complex association behavior. Above all, the $M_{eff}(\rho_o)$-plot is distinctly nonlinear, and its zero crossover is shifted towards those found with apolar solvents. In addition, the slope of $M_{eff}(\rho_o)$ near the zero crossover yields M = 1782 g/mol, which shows that the compound is aggregated in the corresponding solvent mixtures. Aggregation already starts at much lower chloroform contents, and progressively shifts the M_{eff} figures towards higher values with increasing content of this solvent. Thus, the compound's M in acetonitrile cannot be determined using the solvent system acetonitrile/chloroform.

The differences between the present figures for $\overline{v}$ and the figures given in (6) are mainly due to differences in the solvent and to the salt effect. Undetected heterogeneity of the compound's state of association in acetone could be another reason.

According to the data of Figure 4 it seems that, for determining a well-matched pair of M_{eff} and $\overline{v}$, solvents of similar polarity have to be used. More detailed studies on the behavior of other compounds are required before a more general and quantitative description of this condition can be established, guiding the selection of suitable solvent pairs.

Nonideal Sedimentation Behavior: Charge Effects

A general treatment of the sedimentation behavior of large molecules in the analytical ultracentrifuge has to consider several types of nonidealities (9,10). With uncharged molecules, ideal sedimentation behavior (e.g., full validity of eqs. (1) and (2)) can be achieved by sufficiently diluting the sample (which, with the present instrumentation, can be realized easily). With charged molecules, however, both the "primary" and the "secondary" (or "residual") charge effect cannot be eliminated by dilution (9,10). Since the model compound applied in this study as well as many other supramolecular compounds are charged, the charge effects will be considered in the following. To our knowledge, for supramolecular compounds the problems related to the secondary charge effect have not been addressed before.

Primary Charge Effect

It is well-known that, in the absence of salt, sufficiently dilute solutions of homogeneous charged macromolecules yield sedimentation equilibrium profiles which can be perfectly fitted by a single Boltzmann distribution, but with the true M being replaced by $M/(z + 1)$ (where z is the number of ionizable sites on the compound) (9,10). This reduction in the apparent M can be eliminated by adding a "supporting electrolyte", the salt concentration required depending on solute concentration but normally being in the millimolar range (9,10). A demonstration of this effect using a supramolecular compound, a [2x2]-cobalt coordination array, and two different solvents was already given (4). More detailed measurements on the [1x1]-cobalt grid of Figure 1 are shown in Figure 6. It is apparent from the figure that M_{eff} in salt-free solutions is smaller by approx. 45 % than in the presence of sufficient salt, but that the plateau of M_{eff} is reached already at a salt concentration of 1 mM. Thus, the primary charge effect can be as easily eliminated with supramolecular compounds, as with macromolecules.

Secondary (Residual) Charge Effect

The secondary charge effect cannot be easily eliminated. It has attracted much theoretical, but only limited practical interest (9,10,19,20). According to theory, it leads to an apparent reduction in the molar mass, ΔM, which is given by

$$\Delta M = (zM_2/2)\ (1-\bar{v}_2\rho_o)/(1-\bar{v}\rho_o), \tag{3}$$

where z is the number of ionizable sites on one compound molecule and M_2 and $\bar{v}_2$ the molar mass and partial specific volume, respectively, of the supporting (univalent) electrolyte (10). The effect may be regarded as being due to an apparent binding of the added salt on the solute molecules (10). In work on proteins which are not too far from their isoelectric point, the secondary charge effect normally is considered negligible, due to the large difference between M and ΔM. However, Casassa and Eisenberg have shown that it can be taken into account by incorporating the necessary corrections into a specially defined and measured analogue of $\bar{v}$ (10,19,20).

With supramolecular compounds, the situation is different from that with proteins: For the supporting electrolyte used in the present study, NH_4PF_6, M_2 is 163.0 g/mol and $\bar{v}_2$ is 0.402 ml/g (this study). This yields, together with z = 2 and the data given above, $\Delta M = 341$ g/mol, which amounts to 25 % of M and thus strongly exceeds the experimental uncertainty of M. This is true even if one would assume that the undetected presence of dimeric grid would compensate for the reduction of M. In order to demonstrate this, we have reevaluated the A(r)-data applying eq. (2), assuming the presence of both monomers and dimers of unknown apparent monomer mass, M_1^*. We have varied the latter figure, determining the corresponding sum of the squared residuals of the fit, σ. The results are shown in Figure 7, together with the calculated percentage of dimer. The minimum of $\sigma(M_1^*)$ is well-defined and corresponds to virtually zero dimer concentration, and the optimum M_1^*-value agrees, within the limits of error, with the figure for M obtained from the standard evaluation described above. Thus, our data indicate that our measurement yields the true molar mass of the supramolecular compound and leave no room for a correction for the secondary charge effect, as calculated above. It therefore seems to us that, analogous to the Casassa-Eisenberg

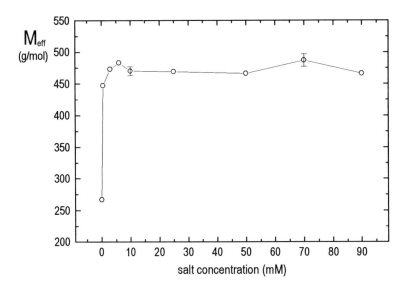

Figure 6. Eliminating the primary charge effect: M_{eff} for the cobalt grid in acetonitrile plus different concentrations of NH_4PF_6, as a function of salt concentration.

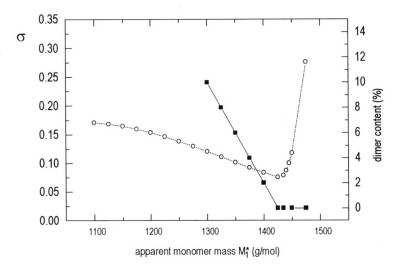

Figure 7. An alternative fit to the data of Fig. 5, based on a monomer/dimer model of self-association: (O) Dependency of the sum of the squared residuals, σ, on the assumed apparent molar mass M_1^* of the monomer (based on $\bar{v} = 0.854$ ml/g). (■) Calculated relative contributions of the dimer (integrated over the sample volume).

approach, our approach for the determination of $\bar{v}$ may already take into account the binding phenomena leading to the secondary charge effect. In part, this may also explain the observed strong dependency of $\bar{v}$ on the solvent.

Discussion

In continuation of our previous work *(4,5)*, the present paper demonstrates that analytical ultracentrifugation, in particular sedimentation equilibrium analysis, is a powerful tool for studying the state of association of supramolecular compounds in solution. The main goal of the present communication was, however, a detailed investigation of two inherent problems of the method. (i) determination of the partial specific volume, $\bar{v}$, of the compound under study and (ii) the so-called "charge effects".

With respect to the $\bar{v}$-problem, we have focussed on the buoyant density method, extending earlier studies *(6,17)*. In contrast to classical densimetry, this method uses the same low range of compound concentrations applied during molar mass determinations in the analytical ultracentrifuge and thus does not suffer from disturbancies due to microcrystal or cluster formation which apparently can occur at much higher compound concentrations *(6)*. Our data clearly show that, in fact, $\bar{v}$ is a quite variable parameter, which for a given compound can vary considerably, depending on the solvent composition. Solvent polarity apparently is a parameter of major influence on $\bar{v}$. At present, it remains unclear whether it is the only or at least the dominating parameter, which would possibly allow predictions of $\bar{v}$-changes in quite different solvents. This topic is now being studied in our laboratory with charged as well as uncharged compounds. At the present stage, the $\bar{v}$-problem remains to be the most difficult aspect in ultracentrifuge studies on supramolecular compounds.

Considering the charge effects, the present study has confirmed that the primary charge effect can easily be suppressed. With respect to the secondary charge effect, our data indicate that the measured molar mass M and partial-specific volume $\bar{v}$ is that of the composite particle, i.e., the monomeric cobalt coordination array plus the two associated PF_6^- counterions. The data could not be modeled with sufficient quality when corrections were applied for the large predicted secondary charge effect *(10)*, even when considering the presence of both monomeric and dimeric cobalt grids. It thus seems that the procedure suggested and applied by us for determining $\bar{v}$, the buoyant density method using two solvents of similar polarity, does not require that the resulting M-values be corrected for the secondary charge effect. The reason may be that, analogous to the Casassa-Eisenberg method *(19,20)*, it is the electroneutral species which is investigated, both in the M- and $\bar{v}$-determinations. In subsequent work, we will try to clarify this aspect as well as to enlarge the experimental data base.

Acknowledgements: We express our gratitude to Professor H. Eisenberg for helpful advice. Our work was supported in part by grants from the Fonds der Chemischen Industrie and the Deutsche Forschungsgemeinschaft (SFB 266 and SCHU 1229/2-1) to U. S. S.

References

1. Lehn, J.-M. *Supramolecular Chemistry - Concepts and Perspectives*; VCH: Weinheim, 1995.

2. Lawrence, D. S.; Jiang, T.; Levett, M. *Chem. Rev.* **1995**, *95*, 2229-2260.

3. Semenov, A.; Spatz, J. P.; Moeller, M.; Lehn, J.-M.; Sell, B.; Schubert, D.; Weidl, C. H.; Schubert, U. S. *Angew. Chem.* **1999**, *111*, 2701-2705, *Angew. Chem. Int. Ed.* **1999**, *38*, 2547-2550.

4. Schubert, D.; van den Broek, J. A.; Sell, B.; Durchschlag, H.; Maechtle, W.; Schubert, U. S.; Lehn, J.-M. *Progr. Colloid Polym. Sci.* **1997**, *107*, 166-171.

5. Schubert, D.; Tziatzios, C.; Schuck, P.; Schubert, U. S. *Chem. Eur. J.* **1999**, *5*, 1377-1383.

6. Tziatzios, C.; Durchschlag, H.; Sell, B.; van den Broek, J. A.; Maechtle, W.; Haase, W.; Lehn, J.-M.; Weidl, C. H.; Eschbaumer, C.; Schubert, D.; Schubert, U. S. *Progr. Colloid Polym. Sci.* **1999**, *113*, 114-120.

7. Schuck, P. *Biophys. J.* **1998**, *75*, 1503-1512.

8. Schuck, P.; Millar, D. B.; *Anal. Biochem.* **1998**, *259*, 48-53.

9. Schachman, H. K. *Ultracentrifugation in Biochemistry;* Academic Press: New York, 1959.

10. Fujita, H. *Foundations of Ultracentrifugal Analysis;* John Wiley: New York, 1975.

11. Schubert, U. S.; Eschbaumer, C.; Weidl, C. H. *Design. Monom. Polym.* **1999**, *2*, 185-198.

12. Schuck, P. *Progr. Colloid Polym. Sci.* **1994**, *94*, 1-13.

13. Schuck, P.; Legrum, B.; Passow, H.; Schubert, D. *Eur. J. Biochem.* **1995**, *230*, 806-812.

14. Durchschlag, H. In *Thermodynamic Data for Biochemistry and Biotechnology;* Hinz, H.-J., Ed.; Springer: Berlin, 1986, pp. 45-128.

15. Durchschlag, H.; Zipper, P. *Progr. Colloid Polym. Sci.* **1994**, *94*, 20-39.

16. Edelstein, S. J.; Schachman, H. K. *J. Biol. Chem.* **1967**, *242*, 306-311.

17. Tziatzios, C.; Durchschlag, H.; González, J. J.; Albertini, E.; Prados, P.; de Mendoza, J.; Eschbaumer, C.; Schubert, U. S.; Schuck, P.; Schubert, D. *Polym. Prepr.* **2000**, *41(1)*, 934-935.

18. Klaerner, P. E. O.; Ende, H. A. In: Polymer Handbook, 2nd edn.; Brandrup, J.; Immergut, E. H.; Eds., Wiley: New York, 1975, pp. IV, 61-113.

19. Eisenberg, H. *Biological Macromolecules and Polyelectrolytes in Solution.* Clarendon Press: Oxford, 1976.

20. Casassa, E. F.; Eisenberg, H. *Adv. Protein Chem.* **1964**, *19*, 287-395.

Chapter 15

Dendritic Methodology Applied to the Prediction, Design, and Synthesis of Sol–Gel Materials

Theodore F. Baumann, Glenn A. Fox[*], and Andrew L. Vance

Lawrence Livermore National Laboratory, 7000 East Avenue, L-091, Livermore, CA 94551

Nanostructured materials can be formed via the sol-gel polymerization of inorganic or organic monomer systems. While sol-gel chemistry offers many opportunities to investigate new material compositions, it suffers from an inability to separate the process of cluster (sol) formation from gelation. This limitation results in structural deficiencies in the gel state that ultimately impact the physical properties of the dried xerogel, aerogel, or nanocomposite. We are utilizing dendrimer methodology to separate the cluster formation step from the gelation process so that new nanostructured materials with controlled topologies can be produced. The dendrimers are used as *pre-formed* clusters of known size that can be cross-linked to form the gel network. Toward this goal, we are synthesizing functionalized dendrimers as gel precursors that allow for greater control over the molecular scaffolding of the resulting gel.

Introduction

During the mid-1980s, sol-gel polymerization of metal alkoxides and selected organic monomers was identified as a method to improve structural control in low-density materials. These new polymerization methods enabled the control of both structure and composition at the nanometer level. For example, a two step polymerization of tetramethoxysilane (TMOS) was developed such that silica aerogels with densities as low as 3 kg/m^3 (~ 2 x the density of air) could be achieved.(1-3) Organic aerogels based upon resorcinol-formaldehyde and melamine-formaldehyde can also be prepared using the sol-gel process.(4-8) Both resorcinol (1,3-dihydroxybenzene) and melamine (1,3,5-triaminotriazine) serve as trifunctional monomers, with resorcinol undergoing electrophilic aromatic substitution with formaldehyde in the 2,4,6 ring positions. These intermediates condense into polymeric "clusters," or sols, ranging in size from 3 to 20 nm. Further reaction of these sol units is believed to involve the peripheral hydroxymethyl moieties that allow these clusters to cross-link and eventually form gels that exhibit a "string of pearls" appearance (Figure 1).

In all cases, sol-gel polymerization depends upon the transformation of monomers into nanometer-sized clusters followed by cross-linking into a 3-dimensional network or gel state. While this technique allows for partial control over properties of the gel materials, it is still an inexact science that has failed to capitalize upon other advances in the fields of polymer science. For example, functionalization of organic aerogels has been, to date, problematic and largely unsuccessful. It is now possible, however, to synthesize organic or organometallic macromolecules of known size, shape and surface functionality.(9, 10) In effect, the step-wise synthesis and isolation of "sol-gel clusters" can be realized through dendritic chemistry. While dendritic molecules were first prepared in 1978, only recently was the first report of a multi-dendritic array published.(11) The preparation of organic-inorganic hybrid xerogels through the sol-gel polymerization of carbosilane dendrimers was also recently demonstrated.(12,13) We believe that the opportunity exists to incorporate functionalized organic dendrimers into the sol-gel process so that materials having tailored composition and high structural efficiency can be prepared(Figure 2).

Structural efficiency refers to the topology or interconnectivity of a 3-dimensional network, gel, foam, or aerogel. Gross *et al.* examined the structural efficiency of resorcinol-formaldehyde (RF) aerogels using beam bending and acoustic techniques. (14) For RF gels synthesized at 10% solids, only 10-15% of the mass is actually interconnected in such a manner that it can support a load, implying that 85-90% of the clusters formed during the sol-gel polymerization are present as "dead ends" or "loops." For RF gels synthesized at 50% solids, the structural efficiency approaches 70%. Structural efficiency is particularly important in applications that require aerogels with good mechanical properties for machining, wetting, or drying. Aerogel processing is also related to the structural integrity of the gels. If the gel is allowed to air dry, the low-

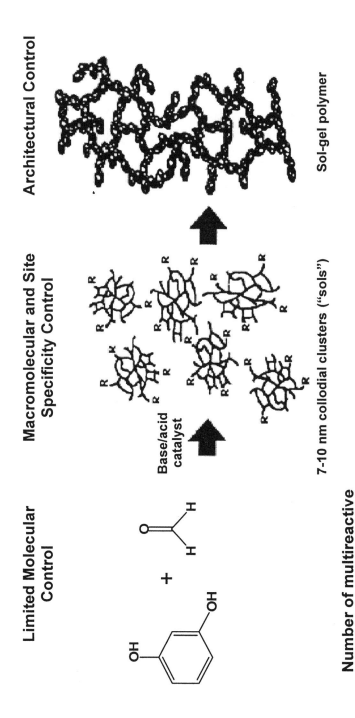

Figure 1. Pathway for traditional organic sol-gel polymerization reactions

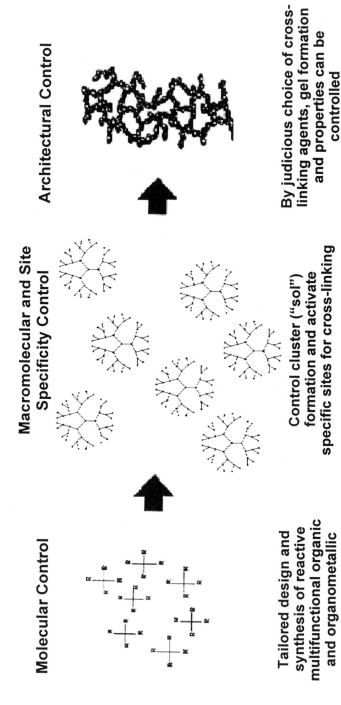

Molecular Control

Macromolecular and Site Specificity Control

Architectural Control

Tailored design and synthesis of reactive multifunctional organic and organometallic monomers

Control cluster ("sol") formation and activate specific sites for cross-linking

By judicious choice of cross-linking agents, gel formation and properties can be controlled

Figure 2. Synthetic scheme depicting the incorporation of dendrimers into sol-gel synthesis

density structure collapses into a high-density mass because large capillary forces are generated when the liquid meniscus moves through the small cells and pores of the precursor. Multiple solvent extractions, followed by supercritical extraction with CO_2, are required to maintain the open cell structure of the gels. Through judicious choice of starting materials and cross-linking agents, we may be able to improve the structural integrity of the resultant gel such that capillary action will not be as destructive to the molecular scaffolding. Thus processing of sol-gel products may not require the lengthy and expensive extraction process.

Dendrigels

As described earlier, the sol-gel process involves the condensation of monomers into higher order clusters that then cross-link to form the three-dimensional gel network. Utilization of dendrimers allows us to separate the cluster formation step from the gelation process so that new nanostructured materials with controlled molecular architecture can be generated. The dendrimers will be used as *pre-formed* clusters of known size that can be cross-linked to form the gel network. With the appropriate choice of the peripheral units, these dendrimers should behave in a manner analogous to the intermediates formed during the sol-gel polymerization.

We initially focused on the PAMAM dendrimers (*15*) as cluster surrogates for three reasons: (1) a range of generations (G_0 to G_{10}) are commercially available, (2) the structure of these dendrimers has been studied and (3) we can exploit the synthetic versatility of the terminal amino groups of the dendrimer to cross-link the dendrimers. Reaction of the peripheral amino groups of the PAMAM dendrimer with appropriate equivalents of formaldehyde, under conditions analogous to those used to prepare melamine-formaldehyde gels (*3*), produces highly cross-linked multi-dendrimer networks. For example, treatment of PAMAM G3 dendrimer (32 terminal amino groups) in 1:1 (v:v) solution of water and MeOH with 16 equivalents of formaldehyde (37% solution in H_2O) affords a stable organic gel (Figure 3). Interestingly, no catalyst was required to initiate the polymerization reaction. The reaction solutions were allowed to gel at room temperature over a 48 h period, after which time, the transparent colorless monoliths were removed intact from the glass vials. In an attempt to obtain an aerogel, the reaction solvent retained by the gel was exchanged for acetone prior to supercritical extraction with carbon dioxide (1200 psi at 50 °C). The structural integrity of the gels is maintained during the solvent exchange, but a significant amount of densification was apparent following the supercritical extraction. The final density of the xerogel monolith was determined to be ~900 mg/cm^3. BET surface area and pore volumes for the xerogel product are currently being investigated.

Several factors, including formaldehyde equivalents, temperature, concentration and solvent, as well as the size of the dendrimer, can be manipulated to influence the architecture of the gel. As a result, we are currently

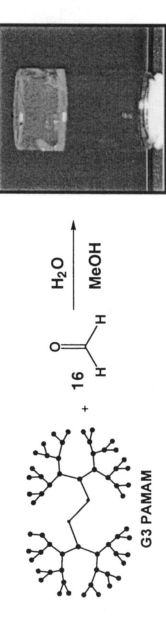

"Dendrigel"

Figure 3. Synthesis of a G3 PAMAM dendrigel with a picture of the gelation product

refining the synthesis of the dendrigels to improve the structural integrity of the materials. We are also investigating the utilization of cross-linking agents other than formaldehyde, such as glutaric dialdehyde and terephthaldicarboxaldehyde. The size and rigidity of these reactants will effect the interconnectivity of the dendrimers and, as a result, the judicious choice of cross-linking agent may allow us to tailor the molecular architecture of the gel.

Functionalized Dendrimers

We are also interested in preparing dendrimers peripherally functionalized with resorcinol units. For example, treatment of a PAMAM dendrimer with 3,5-dihydroxybenzaldehyde should yield a resorcinol-terminated dendrimer that can then be reacted with formaldehyde in the presence of a base catalyst to form an organic sol-gel network. A number of important issues, however, must first be resolved to determine whether such molecules will behave in analogous fashion to the resorcinol-formaldehyde sol clusters. First, we must determine if derivatives of resorcinol retain the reactivity of the parent molecule in the sol-gel polymerization process. Since many of these functionalized dendrimers will not be soluble in water, we also need to find alternative solvent media for the sol-gel reaction. Due to the high cost of dendrimers, the use of smaller, less-expensive model compounds should be extremely beneficial in refining the conditions under which these functionalized dendrimers will react.

Resorcinol Derivatives

Little research has focused on the utilization of resorcinol derivatives in sol-gel polymerization reactions. As stated above, resorcinol undergoes electrophilic aromatic substitution with formaldehyde in the 2,4,6 positions of the ring due to the activating nature of the hydroxyl groups. Functionalization of resorcinol can potentially deactivate the ring and, therefore, severely alter the reactivity of the molecule. We have found that the reactivity of resorcinol derivatives is dependent on the type and position of functional group (Figure 4). For example, reaction of the sodium salt of 2,4-dihydroxybenzoic acid with formaldehyde in H_2O, catalyzed by Na_2CO_3, followed by solvent exchange and supercritical extraction in CO_2, yields a functionalized organic aerogel with a density of 300 mg/cm^3. Reaction of the 3,5-dihydroxybenzoate salt under the same conditions, however, produces a gel that falls apart during the solvent exchange step. Placement of the carboxylic acid group in the *meta*-position (relative to the hydroxyl groups) sufficiently deactivates the ring and limits the amount of cross-linking in the gel. Functionalization of that same *meta*-position with an activating group does not impede the formation of a robust sol-gel product. We

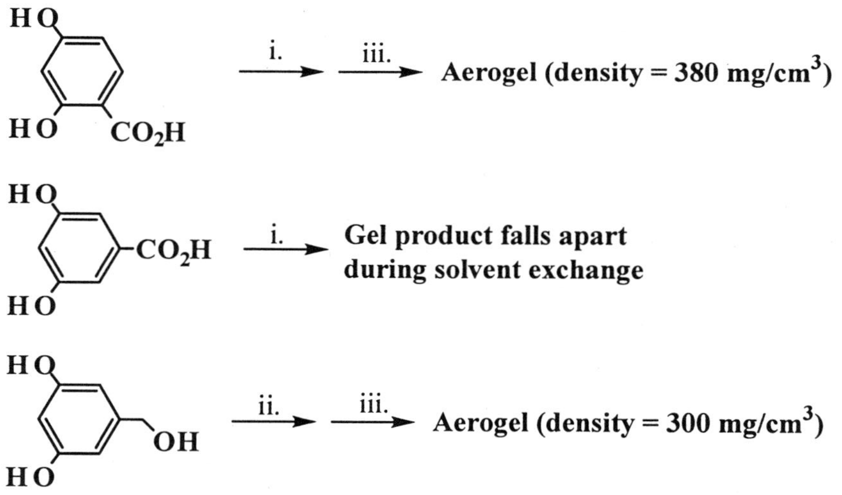

Figure 4. Preparation of organic aerogels from resorcinol derivatives

(i.) 0.5 eq Na$_2$CO$_3$, H$_2$CO, H$_2$0, catalyst; (ii.) H$_2$CO, H$_2$0, catalyst; (iii.) solvent exchange, supercritical extraction.

were able to react 3,5-dihydroxybenzyl alcohol with formaldehyde in H_2O to yield another functionalized organic aerogel with a density of 350 mg/cm^3.

Alternative Sol-Gel Reaction Media

Since many of these functionalized dendrimers may not be soluble in water, alternative solvent systems for the sol-gel reaction are being investigated. The formaldehyde used in this process is a 37% solution in water, so these reactions will never be performed under completely anhydrous conditions. N,N-dimethylformamide was selected as the reaction medium for the preparation of organic aerogels because it is a polar solvent capable of dissolving most organic macromolecules and it is miscible with both water and acetone, two solvents commonly used in the processing of aerogels. The base-catalyzed reaction of resorcinol with formaldehyde in DMF, followed by solvent exchange and supercritical extraction with carbon dioxide, does indeed produce an RF aerogel. The transmission electron micrograph (TEM) of the aerogel shows that the microstructure of these materials are quite similar to those obtained in conventional RF sol-gel reactions, despite the change in reaction media (Figure 5). The resulting aerogels, however, were more dense than the formulation predicted (target density = 150 mg/cm^3, actual density = 300 mg/cm^3). As a result, the conditions for the *non-aqueous* sol-gel polymerization need to be optimized to account for this difference.

Model Compounds

Due to time and cost involved in the synthesis of higher generation dendrimers, we feel that the use of model compounds will be extremely advantageous. These model compounds not only allow us to optimize the sol-gel reaction conditions for the functionalized dendrimers, but they also represent a new class of starting materials for novel sol-gel materials. Since many of the commercially available Starburst dendrimers contain either amine or alcohol termini, we have begun the synthesis of model compounds based on di- or tri-alcohols and amines that containing multiple resorcinol moieties. The synthesis of the model compounds first requires protection of the phenol groups of the resorcinol derivatives. The acetyl group was selected as the protecting group since aryl acetates can be easily prepared from the phenol and acetic anhydride and can be readily cleaved by saponification.(*16*) Since we know that carboxylic acid derivatives are reactive under sol-gel conditions, we have decided to use 3,5-dihydroxybenzoic acid as the starting material for these model compounds. The following procedure (Figure 6) specifically describes the synthesis of compound **3**, but can be used to functionalize a variety of diols and triols.

Treatment of 3,5-dihydroxybenzoic acid with neat acetic anhydride afforded the acetyl-protected product **1** in high yields.(*17*) The diester product **3** was

70.00 nm

Figure 5. Transmission electron micrograph of an organic RF aerogel prepared in N,N-dimethylformamide

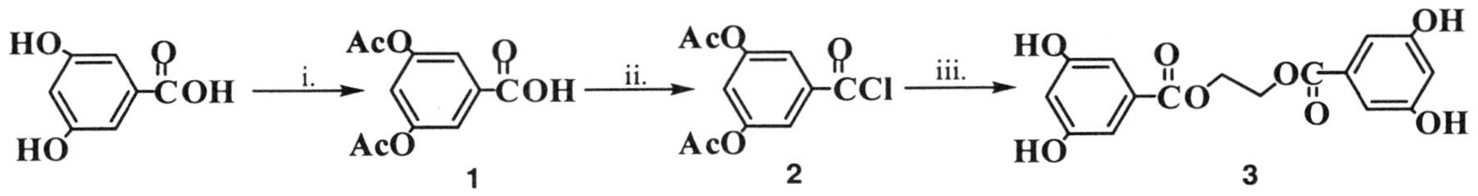

Figure 6. Synthesis of model compound 3

(i.) Acetic Anhydride, 100°C; (ii) SOCl$_2$, DMF, 80°C; (iii.) (a) HOCH$_2$CH$_2$OH, TEA, CH$_2$Cl$_2$ (b) NaHCO$_3$, H$_2$O/EtOH.

prepared through treatment of **1** with thionyl chloride and the acid chloride **2** was quenched with 0.5 equivalents of ethylene glycol. The acetyl groups can then be selectively removed through treatment of acetyl-protected product with a saturated solution of NaHCO$_3$.(*18*) Compound **3**, containing two resorcinol groups, has been reacted with formaldehyde (2 equivalents) in DMF with Na$_2$CO$_3$ as the catalyst. The reaction solution was cured at 80°C for 4 days to give translucent orange gels. During the solvent exchange with acetone, the monoliths did show some cracking, similar to that seen in the gel produced from 3,5-dihydroxybenzoic acid. The physical properties of the aerogel materials, such as BET surface area, pore volume and radius, prepared from compound **3** are currently under investigation. Compound **3** will also serve as the core for the preparation of new dendrimers. For example, divergent growth from the phenol groups of **3** can be used to prepare larger dendritic systems containing peripheral resorcinol moieties.

The method used to prepare compound **3** has also been used to synthesize model compounds in which the resorcinol units are linked by diamines, such as ethylenediamine. Compounds of this type are important not only as models for the functionalized PAMAM dendrimers, but the diamine portion of the molecule can also be used as a ligand for transition metal ions. Gelation of such metal complexes may be a convenient way to incorporate and uniformly distribute inorganic species throughout organic aerogels.

Summary

We are incorporating dendritic chemistry into sol-gel synthesis for the design of new nanostructured organic aerogels. This strategy should provide us with greater control over the molecular architecture of materials formed by the sol-gel process. We are focusing our efforts on the design of functionalized dendrimers that are analogous to the clusters formed in the resorcinol-formaldehyde or melamine-formaldehyde sol-gel reactions. As a first step toward this goal, we have used PAMAM dendrimers as precursors for the synthesis of organic dendrigels. We are currently investigating the influence that the amount and type of cross-linking has on the structure of the dendrigels. We are also interested in the design of dendrimers that contain resorcinol units at their periphery. Functionalization of larger dendrimer systems with resorcinol has been somewhat problematic, so we have prepared smaller compounds containing multiple resorcinol units as models for the functionalized dendrimers. From the synthesis of the model compounds and their subsequent sol-gel polymerization, we have learned that the choice of resorcinol derivatives and gelation media is critical to the synthesis of robust gel product. We are currently using these results as guidelines in our design of new functionalized dendrimers as sol-gel precursors.

This work was performed under the auspices of the U.S. Department of Energy by University of California Lawrence Livermore National Laboratory under contract No. W-7405-Eng-48.

References

1. Hrubesh, L. W.; Tillotson, T. M.; Poco, J. F. In *Chemical Processing of Advanced Materials*; Hench, L.L.; West, J.K. Eds.; John Wiley & Sons, NY, **1992**; pp. 19-27.
2. Tillotson, T. M.; Hrubesh, L.W. *J. Non-Cryst. Solids* **1992**, *145*, 44.
3. LeMay, J. D.; Hopper, R. W.; Hrubesh, L. W.; Pekala, R.W. *MRS Bulletin* **1990**, *15*, 19-45.
4. Pekala, R. W. *J. Mat. Sci.* **1989**, *24*, 3221.
5. Pekala, R. W.; Alviso, C. T.; Lu, X.; Groß, J.; Fricke, J. *J. Non-Crystalline Solids* **1995**, *188*, 34.
6. Pekala, R. W.; Schaefer, D. W. *Macromolecules* **1993**, *26*, 5487.
7. Pekala, R.W.; Alviso, C. T.; Kong, F. M.; Hulsey, S. S. *J. Non-Cryst. Solids* **1992**, *145*, 90.
8. Alviso, C. T.; Pekala, R. W. *Polym. Prpts.* **1991**, *32*, 242.
9. Tomalia, D. A.; Naylor, A. M.; Goddard, W. A. *Angew. Chem. Int. Ed. Engl.* **1990**, *29*, 138.
10. *Dendritic Molecules: Concepts, Syntheses, Perspectives*; Newkome, G. R.; Moorefield, C. N.; Vögtle, F, Eds.; VCH Weinheim, XX, 1996.
11. Balogh, L.; de Leuze-Jallouli, A.; Dvornic, P.; Kunugi, Y.; Blumstein A.; Tomalia, D. A. *Macromolecules* , **1999**, *32*, 1036.
12. Kriesel, J. W.; Tilley, T. D. *Chem. Mater.* **1999**, *11*, 1190-1193.
13. Boury, B.; Corriu, R. J. P.; Nuñez, R. *Chem. Mater.* **1998**, *10*, 1795-1804.
14. Gross, J.; Scherer, G. W.; Alviso, C. T.; Pekala, R. W. *J. Non-Cryst. Solids* **1997**, *211*, 132.
15. Tomalia, D. A.; Hall, M.; Hedstrand, D. M. *J. Am. Chem. Soc.* **1987**, *109*, 1601 and references therein.
16. *Protective Groups in Organic Synthesis*; Greene, T. W.; Wuts, P. G. M., Eds.; John Wiley & Sons, NY, 1999; pp. 276-278.
17. Turner, S. R.; Voit, B. I.; Mourey, T. H. *Macromolecules*, **1993**, *26*, 4617.
18. The acetyl deprotection was a modification of the following procedure: Büchi, G.; Weinreb, S. M. *J. Am. Chem. Soc.* **1971**, *93*, 746.

Author Index

Subject Index

V